MASTERING ERP

A LEADER'S GUIDE TO IMPLEMENTATION SUCCESS

MANOJ THAKUR

For permission, contact: service@wisebytes.us and/or +1 7165-MANOJT.

Mastering ERP
A Leader's Guide to Implementation Success

Published by WISEBYTES SOLUTIONS INC

525 South 4th Street, Suite 585
Philadelphia, PA 19147 USA

wisebytes.us

Cover by WISEBYTES SOLUTIONS INC

ISBN: 979-8-9987962-0-3 (hardback)

DEDICATION

This book is dedicated to my parents, teachers and each member of my personal and professional family with whom I have learned to express my work in the form of this book.

Glossary of Key Terms

Artificial Intelligence (AI) - AI refers to computer systems capable of performing tasks that typically require human intelligence, such as learning, problem-solving, and decision-making.

Application Programming Interface (API) - An API is a set of rules and specifications that software programs can follow to communicate and exchange data with each other.

Automation - The use of technology to perform tasks with reduced human intervention, often used to streamline business processes.

Best Practices - Proven methods or techniques that consistently show superior results and are used as benchmarks.

Business Intelligence (BI) - Technologies and strategies used by enterprises for data analysis of business information to support decision-making.

Change Management - A structured approach to transitioning individuals, teams, and organizations from a current state to a desired future state.

Cloud ERP - ERP software that is hosted on a vendor's cloud platform, accessible via the internet, offering flexibility and lower upfront costs.

Configuration - Adjusting ERP settings and parameters to match business requirements without changing source code.

Customization - Modifying ERP software code to add features or change behavior beyond standard configurations.

Data Migration - The process of transferring data from old systems into the new ERP system, often involving cleaning, mapping, and validation.

Digital Transformation - A strategic process that uses digital technologies to create new, or modify existing, business processes and customer experiences. It involves fundamentally changing how an organization operates by integrating digital technologies into all aspects of its business.

Go-Live - The point in time when an ERP system becomes operational and is used in a live production environment.

Implementation Partner - A third-party consultancy or service provider that supports the implementation of a new system or software.

Integration - The process of connecting ERP systems with other software or data sources to ensure seamless information flow.

Key Performance Indicator (KPI) - Measurable values that demonstrate how effectively a company is achieving business objectives.

Legacy System - An outdated information system that may still be in use but is being replaced or integrated into a new ERP.

Module - A distinct component of an ERP system that focuses on a specific business function (e.g., finance, inventory, production).

On-Premises ERP - ERP software installed locally on a company's hardware and servers, rather than in the cloud.

Scalability - The ability of an ERP system to grow and handle increased demand as the business expands.

Stakeholder - Any individual or group that has an interest in or is affected by the ERP implementation, including executives, employees, and vendors.

System Integrator - A company or individual that ensures ERP systems work seamlessly with other IT infrastructure.

MASTERING ERP: A LEADER'S GUIDE TO IMPLEMENTATION SUCCESS

Are you facing the daunting challenge of implementing an enterprise resource planning (ERP) system that integrates with your business operations? In a complex commercial climate, the high failure rate for an ERP project may cost your company millions of dollars. Yet, the promise of a fully integrated system—one that streamlines processes, improves data accuracy and sharpens strategic decision-making—remains irresistible.

This book presents a no-nonsense, step-by-step guide based on decades of experience, transforming an almost insurmountable challenge into a clear, actionable roadmap for success.

Real-World Experience

By drawing on real-world experiences from small and mid-sized businesses to global enterprises, this book gives practical insights about choosing the right technology partner, balancing customization with standardization and building a culture that bridges the gap between technical solutions and everyday operations. Each chapter unveils strategies for meticulous planning, efficient execution, overcoming resistance and harnessing an ERP solution to fuel long-term growth and operational excellence.

C-Level Guidance

From CEOs to CFOs and their operational managers, business leaders will discover how to transform ERP investments into strategic advantages that deliver measurable improvements and a lasting competitive edge. This book is your essential companion—a compass guiding you through the pitfalls and triumphs of ERP implementation—ensuring you make informed, confident decisions for every step.

Successful Results

Don't let your project become another negative statistic. It's time to take control of your future and transform your organization into an agile, data-driven powerhouse. In these pages, uncover practical, fail-proof solutions and start your progress toward an ERP system that works to propel your business to new levels of success. Take the steps needed to secure your organization's competitive advantage.

Introduction

When you receive a report indicating that your ERP project is 60% over budget, it's more than just a difficult day. It's a wake-up call highlighting the high-stakes, high-risk world of enterprise resource planning system implementation. Gartner research says that over 70% of ERP projects fail to deliver their intended results. The issue isn't just technical missteps. It's the intricate balancing act between strategy, technology and, most importantly, people.

Over 25 years of working in this field have shown me one undeniable truth: ERP is not just a software upgrade. It's a full-scale business transformation. Success depends on aligning systems, processes and human judgment to drive lasting change.

Practitioner's View

This book isn't a theoretical manual. It's a collection of honest accounts detailing challenges, mistakes, lessons learned and victories that real organizations have faced during their ERP journeys. The goal is to instill confidence in managing your ERP implementation amid uncertainty, setbacks and changes that are often unpredictable and can arise at any moment.

ERP systems are not merely about technology. They establish a resilient operational foundation that supports growth and adaptation in an ever-evolving business landscape. Throughout these pages, you will discover actionable strategies for managing complexity, making informed decisions and implementing an ERP system that serves as the backbone of your organization.

By providing practical insights, this book aims to empower you to navigate the intricacies of ERP with greater confidence and clarity, ensuring that you are well-equipped to tackle the hurdles that may lie ahead. Each story shared within these pages reflects the diverse realities faced by different companies, illustrating that while the journey may be fraught with difficulties, it also offers incredible opportunities for development and success.

What This Book Covers

This book is your real-world guide to ERP success—a clear, action-oriented roadmap grounded in experience. It offers lessons learned from boardrooms, strategic planning sessions, IT initiatives and functional operations, including remarkable successes and painful failures. Rather than presenting academic theory, it delivers practical insights and tactical advice for leaders managing ERP projects.

You'll find answers to critical questions such as:

- How do you evaluate whether an ERP system is right for you?
- How do you choose an implementation partner whom you can trust?
- How do you integrate the human element into a technology project?
- How do you produce tangible, lasting results?
- How do you avoid recurring pitfalls during implementation?
- How do you assess whether the ERP implementation was a success?
- How can you sustain ERP success for your enterprise?
- What does tomorrow look like for ERP and its implementation?

About the Author

Manoj Thakur

With over 25 years of experience implementing ERP solutions for manufacturing and distribution businesses, including Fortune 100 global enterprises, Manoj Thakur has witnessed firsthand what leads to successful or failed ERP implementations. His straightforward and tailored approach has helped many projects realign and succeed. Beyond having a master's in business administration (MBA), his continuous learning efforts earned him the following certifications from prestigious professional institutions.

- Digital Transformation from Cornell University
- Digital Marketing from Columbia University
- Project Management Professional from PMI
- Supply Chain Professional; Transformation of Supply Chain; Logistics, Transportation and Distribution; and Supply Chain Operations Reference (SCOR) Professional Endorsement from ASCM
- APICS Supply Chain Instructor
- AI Engineer Master's Program from SimpliLearn
- Higher Diploma in Software Engineering from Aptech

He passionately practices the lessons learned from consulting engagements, research and his interactions with respected professionals, teachers, instructors and colleagues.

His career exemplifies how using a distilled 20% of the knowledge can be applied to all the tasks at hand without getting distracted by the remaining 80% of the not-so-effective portion in the respective body of knowledge. When searching for any unfathomable answer, he reverently turns the pages of the Sri Bhagavad Gita and submits to his role model, Sri Krishna.

When he is not consulting, Manoj enjoys reading and spending quality time with his lovely wife, daughter, daughter-in-law and son. As an avid sports lover, he never misses key tournament matches in cricket, football and tennis. At his prime in sports, he led the Boulder Cricket Club to many victories and won the recognition "Best Bowler in T20 League" from the Colorado Cricket League. Occasionally, he plays tennis and ping-pong. He listens to music, learns from spiritual discourses and practices Tabla (percussion musical instrument) lessons to nurture his creativity.

To nurture your curiosity and seek advice, stay in touch with Manoj via:

LinkedIn: www.linkedin.com/in/erpcoach

Facebook: www.facebook.com/erpcoach1

Preface

What is ERP?

Enterprise Resource Planning (ERP) refers to business software that integrates and manages core business processes across various departments within an organization. An ERP system serves as a centralized platform that collects, stores and analyzes data from different business functions, including finance, distribution, supply chain management, manufacturing and customer service.

A Single Source of Truth

One of the fundamental aspects of ERP systems is the way they consolidate disparate data streams into a single source of truth. In organizations where multiple departments perceive their data as proprietary, aligning everything in one system can feel like asking them to give up a part of their identity.

Nevertheless, I have consistently observed that when data is centralized, decisions are made faster and with greater confidence. I recall a particularly striking episode with a medium-sized enterprise that had long struggled with budget overruns and delayed reporting. The data was scattered across the organization, and each department used its own methods of tracking performance, making it nearly impossible for top management to get an accurate overview.

Once they agreed on a unified ERP system, the transformation was dramatic. Suddenly, the CFO was no longer chasing numbers from different spreadsheets; instead, a single platform had a clear, real-time view of the company's financial health. This reduced the margin of error and instilled a sense of shared accountability across the organization.

ERP Implementation Challenges

Most of the implementation challenges stem from how ERP software solutions are sold, purchased and implemented. Software publishers, reseller partners and businesses implementing ERP solutions all play their part in these processes. At times, it is a lack of awareness. Other times it is a low-cost offer. Most of the time, it is unvetted influences from a software publisher, their representatives or their marketing efforts.

A small investment in having a working prototype prior to finalizing an ERP system could enhance the return on investment (ROI) and reduce the risk of failure significantly. The on-demand availability of cloud computing infrastructure makes this prototyping effort very doable.

One strategy is to replace the antiquated cycles of "requirement gathering and demo" by inviting ERP users to beta test a prototype of a new ERP system. Let them experience how the new ERP system will help them perform their daily tasks/ They can evaluate the solution for each daily task performed using the prototype and their own data, processes and business rules.

Likewise, business leaders could experience the ERP solution, as a prototype, before executing a long-term contract.

ERP software is built using the most common business practices across a targeted industry and/or targeted business processes. Any practice or business process that was not in such a targeted scope may still be part of the ERP software, but with limitations. However, ERP software's marketing practices often blur such foundational attributes. That makes it harder for a business leader to pick an ERP system that is satisfactory for their business practices. The challenge of inefficient last-mile delivery is relevant for ERP software as well. Even the closest-fitting ERP software may still not have the required depth to support a few critical business practices.

The following are some specific examples of situations and challenges faced during ERP implementation.

Commonly Over Budget

No one wants to be the CFO who, months after launching an ERP project, finds the project over budget and several months behind schedule. Faced with the daunting decision of either abandoning the project or approving an additional 60% in spending, this leader was caught in a maelstrom of panic and uncertainty.

I was then brought in to manage the problems, directly addressing these fears by sitting down with both the CFO and the CEO. We discussed the project expectations, questioned long-held assumptions and recharted a course that later resulted in a positive turnaround. The project was completed for less than the additional cost threshold and with functionalities that went well beyond the initially desired scope.

Stories like this underscore the fact that success in ERP implementation relies heavily on the ability to reassess and recalibrate, to understand what isn't working. It also highlights the limitations of estimating an ERP implementation initiative. Most elements are constantly changing, except for the seemingly magical certainty of the budget based on one's experience or a consultant's wisdom from previous projects. Are there any two identical ERP implementations anywhere? Probably not.

Misaligned Expectations

I remember one instance when working with a mid-size manufacturer. The project, which had initially been expected to serve as a routine update of an existing system, soon revealed complexities that no one had fully acknowledged. The team found themselves struck by misaligned expectations and unforeseen technical challenges. It was a turning point where every decision made had far-reaching consequences. This situation, like many others, taught me that the foundation of any successful ERP venture is a deep comprehension of both technology and the business environment it supports.

Fitting a Square Peg in a Round Hole

Consider the case of a vendor executive who was championing the solution by customizing an ERP meant for a vastly different industry than that of the client. In his hands, the promise of the solution thus crafted appeared alluring, but my analysis pointed to a grim future: the solution fitment was so low that stakeholders would quickly lose faith.

Recognizing the warning signs, I withdrew from the project - an act that proved wise when the entire initiative was later discontinued and the vendor was shown the door. This reminds us that quick fixes and one-size-fits-all solutions rarely deliver lasting value. Instead, true success comes from a willingness to step back, critically examine the situation and make tough decisions that protect the organization's interests in the long run.

Lack of Vendor Alignment

A deep and genuine understanding of one's business needs, coupled with a skeptical but open-minded view toward vendor claims, can save considerable time and resources. An ERP vendor or their reseller partner is undoubtedly capable of demonstrating the ERP's features and functionalities for the business conditions that were assumed to be true while developing the ERP software. How valid are those assumptions for the customer's business, and where is the ERP system going to be implemented? If the vendor does not appreciate this requirement and is unwilling to work with the implementation team, then alignment does not exist.

For many leaders, the lure of an ERP system comes from its promises of streamlined operations and instant access to data. However, reality often involves an intricate path of aligning project goals with organizational culture, connecting disparate technologies and managing expectations that might not always match the capabilities of the selected system.

ERP implementation becomes a technical project and an organizational transformation - an undertaking that tests the resilience and vision of leaders at every level. The challenge is seeing the broader view, so every decision contributes to cohesive, long-lasting improvements.

Rushed User Enablement

There have been instances when the implementation team underestimated the necessity of training, and the user enablement phase was rushed to avoid changing the "go-live" date. Technology might be state-of-the-art, but without the human component equipped to use it effectively, even the best system can falter.

In one project, a lack of proper user training led to the underutilization of several important system features. That experience underscored how essential it is to allocate sufficient time, resources and energy toward supporting ERP system users from day one.

Assuming User Acceptance

I recall assisting a CFO who found himself at a crossroads - facing a project that had ballooned in cost and timeline. That moment of crisis forced him to reconsider conventional strategies and reaffirmed the necessity of continuously questioning project assumptions. Operational team members had different opinions about the outcomes of the implementation. Hence, their contributions to the implementation were less than expected. A series of transparent communications highlighting the "to-be" state of operations improved the buy-ins.

Later, the project stayed within a revised budget and delivered additional functionalities that were never originally planned. These experiences are precisely the kind of stories I want to share with you here. They serve as reminders that even the most daunting setbacks can be overcome with the right mix of informed decision-making, strategic vision and courageous leadership.

In another situation, the initial proposal of an ERP Solution met with resistance, not because the concept was flawed, but because it did not seem to reflect the daily realities of the business. By bridging that gap—ensuring that the new system supports, rather than disrupts, established workflows—leaders can build a valid business case and secure the buy-in necessary to drive the project to completion.

In sharing these accounts and insights, I intend to present real-world advice that goes beyond theory. I have seen all too often how well-intentioned projects have derailed when leadership loses sight of the strategic objectives.

TABLE OF CONTENTS

ERP

Chapter 1: Understanding ERP Systems

Earlier operational systems, like material requirements planning (MRP), focused on inventory and production scheduling. Improvements with MRP II expanded this effort to encompass broader manufacturing operations.

ERP emerged as a business management solution because it could integrate various business functions, such as finance, distribution, manufacturing and supply chain management. ERP streamlines processes, improves data flow and enhances decision-making across an entire organization.

Subsequent ERP systems extended integration to include external partners and stakeholders, leveraging e-business technologies for enhanced collaboration and supply chain management.

This reflects the increasing interconnectedness of modern businesses and the need for real-time data sharing. Since their inception, ERP systems have been continuously adapted to meet evolving business demands and now have taken the shape of a "business platform" that facilitates the launch of spirited initiatives in the multi-dimensional growth of an enterprise.

The widespread implementation of enterprise resource planning (ERP) systems reflects technological advancements in response to shifting business needs. Modern ERP systems incorporate technologies like blockchain for enhanced transparency and security, and artificial intelligence (AI) for improved decision-making and automation.

The Evolution of ERP: Concept to Necessity

The story of ERP systems began long before boards of directors started discussing digital strategy. Years ago, companies struggled to manage their various operations using separate, isolated software tools. J.I. Case, a tractor manufacturer, worked with IBM in the 1960s on ERP. After initial interest expressed by enterprises, mainly manufacturers, demand grew for a solution that could neatly tie together disparate business functions.

Over the years, ERP systems evolved from rigid, single-purpose applications to dynamic platforms that adapt to industry needs. There was a time when the idea of an all-encompassing system seemed like a lofty dream reserved for larger enterprises. In those days, most companies settled for piecemeal solutions that barely scratched the surface of their operational challenges. Each business unit worked in isolation, using separate tools that rarely communicated with each other.

The arrival of ERP systems redefined that mindset. Instead of being seen as a monolithic IT project, ERP started to acquire an identity as a comprehensive business solution—a platform that supported daily operations and actively contributed to strategic decision-making.

The evolution of these systems was not linear. Early efforts focused on integration but were hampered by technological limitations and organizational resistance. Many companies implemented systems that promised to align operations but stumbled when the reality of change set in.

I read about a project in which teams grappled with a clumsy interface and disconnected data sources. The implementation was technically sound on paper, yet it created more confusion than clarity in practice. Stories like these, though difficult in the moment, became powerful lessons that gradually refined the concept of ERP.

In recounting the evolution of ERP systems, I aim to present more than a chronology of events. I share these insights with the hope that they spark a sense of possibility and determination among those who face the daunting task of transforming operations.

Let the history of ERP evolution remind you that every meaningful change starts with a willingness to acknowledge what isn't working and the courage to try a different approach. Through both failure and success, the path toward integrating a truly functional ERP solution is marked by constant learning and a commitment to aligning technology with the heart of your business.

From Disparate Systems to a Unified ERP Framework

From an objective standpoint, the progress from disparate systems to a unified ERP framework represents a microcosm of modern business evolution. Companies have progressively recognized that the information silos of the past produce inefficiencies that are unacceptable in an era defined by rapid change.

I have witnessed a shift from fragmented programs to integrated solutions, a change that has transformed operational processes and underlying competitive strategies. The integrated approach has enabled organizations to optimize resource allocation, streamline processes and make decisions based on accurate, real-time information about their operations.

These improvements saved valuable time and money and helped businesses manage the complexities of modern markets with greater confidence.

A Shift in Design Approach

The shift in design approach came as organizations started to understand that technology is only one part of the equation. ERP systems grew to reflect a more balanced approach that considered business processes, human interactions and cultural dynamics.

During one engagement, I witnessed the transformation of a company that had traditionally managed operations through ad-hoc solutions. The leadership decided to invest in an ERP solution that required a thorough rethinking of their processes. This move initially resulted in challenges, particularly in aligning the new system with established routines. However, sustained conversations with every department and direct involvement in problem-solving eventually made the new way of working acceptable and preferable.

The project underscored an essential insight: technology alone cannot solve all problems. Instead, integrative planning and behavioral insights eventually pave the way for a successful transformation.

ERP Born Out of Necessity

The origin of ERP systems is inextricably linked with the need to drive operational efficiency on all levels of an organization. In many ways, ERP was born out of necessity—a response to the chaotic state of isolated systems that left many companies scrambling for coherence.

The drive for operational clarity has only grown more critical as markets have become increasingly competitive. In a business environment where the margin for error is razor-thin, being able to see the entire process is an imperative. I have seen how organizations that adopt integrated systems gain an advantage by developing a deeper understanding of their process flows and by making faster, more informed decisions.

The Transformation of Vision into Reality

The trajectory of ERP systems mirrors the application of technological progress in business. Early adopters confronted immense challenges in translating their vision into a reality that practical day-to-day work could sustain. Those pioneers often encountered resistance from within, as employees and management alike questioned the value of investing heavily in a new way of doing business.

The senior employees' notorious cynicism was born from past experiences where high hopes were dashed by underperforming systems. It was in these moments of pressure and doubt that I learned one of the most valuable lessons: the success of an ERP implementation depends not just on the software itself but on the willingness of everyone involved to adjust and commit to a transformed way of working.

Technologies Leading the Way

Another dimension to consider is the evolution of technologies. With each new ERP version, the interface has become more user-friendly and adaptive. Modern ERP platforms include real-time analytics, enhanced security features and flexibility to accommodate business priorities.

Over time, what once required extensive customization with deep technical support has become a standard from leading vendors. The advancements of low-code and no-code capabilities simplify the extensibility of ERP systems that can be managed by internal teams.

I have witnessed the shift from systems that demanded relentless troubleshooting to more intuitive solutions that anticipate user needs. This progression has significantly lowered the barriers to entry, enabling even mid-sized businesses to compete at a level once reserved for larger enterprises.

Seeing the Broader View

This narrative is built on the belief that progress in business is achieved not by clinging to old habits but by challenging ourselves to see the broader and contemporary view. By contemplating the origins and subsequent growth of ERP systems, you can better appreciate the complexities and opportunities inherent in modern operational frameworks.

Armed with this context and a clear understanding of both the technological and human factors at play, every decision you make during your ERP implementation will carry the weight of insight gathered from decades of hands-on experience.

In the end, the evolution of ERP stands as both a challenge and an invitation. It calls for a reexamination of current practices with an open mind and readiness to adopt solutions that truly align with business needs. The continual process is one where reflective learning and deliberate action intersect, guiding organizations toward a future defined by operational clarity, informed strategy and sustainable growth.

Use Cases Propel the Evolution

ERP evolution continues chasing real-world needs, use cases, measured adjustments and continuous optimization. It is a process where every error becomes a learning moment, and every breakthrough reinforces the conviction that with the right approach, even the most complex systems can be turned into reliable growth engines. These use cases, which in the past were considered to have a low probability of occurrence, now, in the modern business environment, create opportunities, and whoever gets that right leads the pack.

This continuous evolution is a reminder that ERP systems, like many other business tools, do not create a universal solution. They must be adapted to respond to each organization's specific challenges and opportunities.

There have been instances where organizations attempted to force a one-size-fits-all solution without sufficiently considering business needs, only to encounter friction and disappointing outcomes. These experiences serve as a reminder that while ERP platforms create a framework, the real value lies in thoughtful configuration to fit business realities.

I have seen clients achieve remarkable operational turnarounds when their ERP solutions were tailored to their operational nuances—even when that meant making tough decisions about process redesign or technical tweaks. Each tailored solution reaffirmed the understanding that technology is an aid to human strategy and not a substitute for it.

By understanding this evolutionary nature of ERP, you are better prepared to guide your organization through the challenges of modern business, while having the confidence that technology supports rather than complicates your strategic goals.

Key Components of an ERP System

An ERP system usually has configurable control centers where decisions and processes flow into a unified system. This is the promise and reality of a well-designed ERP system that offers adaptability.

Having witnessed how the appropriate integration of technology and organizational processes can transform operations, I have seen that an effective ERP solution is far more than a mere collection of software modules. ERP operates as the nervous system of a business, connecting various functions and ensuring that data, operations and insights come together to drive smarter decisions.

In my interactions with business leaders, one recurring theme has been the challenge of understanding how these pieces fit together. As the complexity of modern enterprises grows, so does the need for a well-thought-out system that aligns with every facet of the organization.

Architectural Framework

Beyond the operational and human challenges lies the architectural framework that forms the backbone of an ERP system. The modern ERP environment is built to be adaptive. Gone are the days when systems were rigid and could only handle a narrow range of functions. Contemporary platforms are designed with modularity in mind.

An ERP system follows an architectural framework that ensures seamless integration of business processes, data management and scalability.

The key components of an ERP architectural framework include the following elements.

Business Architecture

Defines the business processes, workflows and operational goals that the ERP system supports. A flexible business architecture ensures that ERP aligns with the strategic objectives of various business operations models. Many ERP failures stem from the misalignment between business needs and system capabilities.

Application Architecture

Structures the ERP software components, including modules for finance, HR, supply chain and manufacturing. Modular ERP systems allow businesses to scale efficiently. Companies should prioritize interoperability to avoid vendor lock-in.

Data Architecture

Manages data storage, retrieval and security, ensuring consistency across different departments. Data governance is critical—poor data management leads to inefficiencies. A centralized data model prevents duplication and enhances reporting accuracy.

Technology Architecture

Covers the underlying infrastructure and technological prowess to meet the demand of modern business operations.

1. **Three-Tier Architecture**

   Divides ERP into three layers—database, application and presentation—ensuring scalability and security. Best suited for large enterprises that require high performance and security. It enables businesses to scale operations without overloading a single system.

2. **Service-Oriented Architecture (SOA)**

Uses modular services that communicate via APIs, making integration with third-party applications seamless, which is a game-changer for businesses needing flexibility. Allows companies to integrate CRM, e-commerce and analytics tools without disrupting core ERP functions.

3. **Microservices-Based ERP**

Breaks ERP functions into independent services that communicate via APIs. Highly scalable and ideal for tech-driven enterprises. It allows businesses to update individual services without disrupting the entire system.

4. **Multicloud ERP**

Integrates multiple cloud providers to enhance scalability and redundancy. Great for global enterprises that need high availability and disaster recovery. However, managing multiple cloud environments requires strong IT governance.

5. **AI-by-the-Side Architecture**

This is an upcoming design paradigm that intends to harness the capabilities of artificial intelligence (AI) and offers agentic performance with near-real-user experience for optimal outcomes. AI helps users orchestrate their thoughts, decisions and productivity.

Integration Framework

Ensures connectivity between ERP modules and external systems through APIs and middleware. API-first architecture enables seamless integration with third-party applications. Businesses should prioritize real-time data exchange for better decision-making.

Security and Compliance

Implements authentication, authorization and regulatory compliance measures, such as:

1. **Role-Based Access Control (RBAC)** – Enhances security by limiting user permissions.
2. **Regulatory Compliance Features** – Ensures adherence to industry standards like GDPR, SOX or HIPAA.

Useability Framework

From a usability perspective, ERP architecture offers preferences that are suitable according to business size, industry and future scalability.

The most common preferences considered while selecting an ERP system are the following.

Deployment Model

- **Hub and Spoke** – A setup where corporate headquarters use a primary ERP system, while subsidiaries or business units operate a separate, often lighter ERP. This is ideal for fast-growing companies with multiple locations. It allows subsidiaries to use a cost-effective ERP while maintaining integration with the central system.

- **Cloud** – Ideal for businesses seeking flexibility, remote access and minimal infrastructure management. Hosted on external cloud servers, offering accessibility and reduced infrastructure costs. Perfect for SMBs looking for low upfront costs and easy scalability. However, businesses must ensure strong cybersecurity measures to protect sensitive data.

- **On-Premise** – Offers greater control over data security but requires substantial IT resources.

- **Hybrid** – A mix of cloud and on-premise solutions for businesses needing a balance between security and accessibility. Best for industries with strict compliance needs, such as finance and healthcare, where some data must remain on-premise while leveraging cloud benefits.

Modular Vs Monolithic

- **Modular ERP** – Allows businesses to implement only the required functionalities, improving customization.

- **Monolithic ERP** – Features all functionalities in a single system, often more rigid but with stronger integration.

Integration Capabilities

- **API-First Architecture** – Enables seamless integration with external applications like CRM, e-commerce and third-party tools.

- **Microservices-Based Architecture** – Ensures scalability, breaking ERP functions into independent services.

Scalability and Performance

- **Multi-Tenant Cloud ERP** – Cost-effective and scalable, but shared resources can affect performance.

- **Single-Tenant ERP** – Dedicated infrastructure improves performance but is costlier.

User Experience – Click vs. Converse

Defines how users interact with the ERP system, ensuring ease of use and accessibility. Intuitive UI reduces training time and improves adoption rates. Businesses should focus on mobile-friendly ERP interfaces for better accessibility.

With the AI-by-the-Side Architecture, users' preferences will expect a conversational interface. No more scanning through menu options or clicking the tiles. Just speak in plain business language through any secured device to interact with the ERP system and get things done.

Human capabilities should be leveraged for decision-making where they belong. Let the data entry, order taking, transaction processing and error resolution be handled by the AI, while sitting by the side.

The Core of ERP Implementation

Understanding ERP systems brings us to a point where the core of ERP implementation emerges as a blend of technological advancement, process integrations and human endeavors. Over the course of this exploration, I share experiences that illustrate the transformational potential of ERP when it is tailored and executed with precision and embedded flexibility to manage unknowns. This section ties together the recurring themes of unified vision, data centralization and the challenges and rewards of aligning diverse operational elements under one system.

Reflecting on the evolution of ERP systems, it becomes apparent that successful implementation is not solely about adopting a new technology. It is about rethinking existing business processes and creating a framework that responds to the nuanced needs of organizations.

This transformation begins with a clear understanding of what each component of an ERP system represents. ERP systems streamline decision-making, improve communication and transform raw data into actionable business insights.

One striking lesson drawn from years of observation is that a well-executed ERP implementation does more than integrate the functions of finance, the supply chain and customer service. It creates a uniform flow of information across the organization, one that reduces data silos and time-consuming manual reconciliations.

I have often observed how businesses, previously mired in a maze of spreadsheets and fragmented data sets, suddenly gain a unified perspective that unlocks new levels of operational efficiency. When all teams—from the shop floor to the boardroom—have access to consistent, real-time information, decisions are made with greater accuracy and confidence. This shared understanding fosters a spirit of collaboration that can turn isolated improvements into sustained competitive advantages.

Strategic Alignment

Implementing an ERP system is, above all, an exercise in strategic alignment. It demands that every login, click, action, conversation and automated alert be calibrated to support the broader business strategy.

I have often told executives that technology is a facilitator rather than the end goal. The aim is to create a system that captures operational data and helps drive strategic decisions that propel the company forward. In boardroom discussions, the conversation frequently shifts from the mechanics of the system to how the insights derived can be used to improve competitive positioning, streamline costs or identify new market opportunities.

In practice, this means that the ERP becomes an indispensable tool in strategic planning and a repository of knowledge as dynamic as the market itself.

Cross-Functional Clarity

As you reflect on ERP systems' cross-functional clarity, consider how each element—from data centralization and process integration to user experience and strategic alignment—contributes to unify enterprise functions, building an operational powerhouse. It is not a single component that makes these systems successful, but synergy among all parts. The value lies in the way these components come together to create clarity amid complexity, turning raw data into decisions that align with long-term business objectives.

Over time, I learned that the transformation yielded by a well-implemented ERP system is often gradual. It requires a patient, measured approach—one that acknowledges the current landscape while laying the groundwork for future improvements.

By monitoring system performance, investing in user training and maintaining a dialogue between technological prowess and leadership commitment, organizations can achieve a level of operational clarity that has far-reaching implications. Every improvement, every adjustment, builds upon itself to form a comprehensive solution that supports the entire ecosystem of the business.

The more you investigate the intricacies of ERP systems, the more you may appreciate the delicate balance that holds everything together. It is a balance between technological rigor and human perception, between standardization and customization and between immediate needs and future potential.

This balance is not easily achieved; it must be meticulously cultivated through careful planning, active management and a commitment to continuous improvement. From early days spent wrestling with fragmented data to witnessing full-scale transformations that redefined operational frameworks, the lessons learned have been both varied and invaluable.

Business leaders who understand these nuances are better equipped to make decisions that protect and enhance the organization's operational backbone. Every challenge encountered along the way—from data silos to user resistance—serves as a lesson in what to avoid and, more importantly, what to pursue.

The underlying theme is clear: investing in the thorough understanding and strategic alignment of your ERP system pays dividends that go beyond financial outcomes. It shapes the very way your company operates, sets the tone for employee engagement and lays the foundation for sustained competitive advantage.

Intelligent and Interdependent Integration

Through these stories and insights, a consistent thread emerges: Success in ERP implementation is achieved through a balanced approach. One must consider the technical framework and the human, cultural, and process factors that come into play.

I have spent countless hours with executives grappling with questions of future readiness, balancing the bumpy early days with the long-term competitive gains that result from a well-integrated ERP solution.

Whether rethinking finance processes, restructuring supply chains or redefining customer engagement strategies, the guiding principle was always clear: intelligent integration yields transformative outcomes.

The evolution of ERP systems from a concept to a critical business necessity stands as a testament to the power of strategic planning. The process encapsulates lessons of perseverance, adaptability and the value of maintaining a clear focus on business objectives.

The interdependent nature of an ERP system implies that every change and adjustment matter across the enterprise's processes. The process of full ERP integration includes thoughtful decisions, collaborative planning and an unwavering focus on aligning technology with a business strategy.

The insights shared here aim to create a deeper understanding of what makes an ERP system truly effective—a unified solution that brings clarity to complexity and transforms data into strategic insight. An ERP system's true strength lies not in isolated functionalities but in the cohesive, strategic framework it creates—a framework that is as resilient and adaptable as the business it supports.

Leaders who recognize the importance of integration are better positioned to overcome internal resistance, overcome unforeseen hurdles and secure lasting benefits for their organizations. The lessons captured here are drawn from a career filled with challenging projects and rewarding turnarounds—a blend of hard data, lived experiences and the conviction that careful planning, backed by careful execution, can drive even the most complicated ERP projects toward success.

Meant For Humans

Implementing ERP systems is not about replacing humans or their judgment with automated algorithms and artificial intelligence but about creating environments where technology and human insight work in tandem for the benefit of the enterprise.

The best technology supports people rather than constrains them. In every project I have been involved with, the conversation always centered on how the new system could be molded to support human decision-making and operational creativity. That connection between thoughtful leadership and state-of-the-art technology has been the driving force behind the success stories shared over the years.

Resistance to Change

Merging the various components of a business into one coherent framework is fraught with challenges. Often, the greatest obstacle is not the technology itself but rather the human factor, with its inherent resistance to change. When individuals become accustomed to certain routines, asking them to adopt new ways of working can trigger fear of the unknown.

I sat with teams who spent hours debating the merits of their tried-and-tested methods, only to realize later that their reluctance was less about the system's functionality and more about the perceived threat to their established roles. Addressing these concerns requires patience and candid dialogue. It also entails demonstrating—not just telling—how the new system will make their daily activities less cumbersome and more rewarding.

Facilitating the changes from within each of the employees is often a successful way forward. Humans are motivated by their level of acceptance of the changes. Any change that is thrown at them without earning their buy-in results mostly in failure.

Here, communication becomes the cornerstone of such facilitation, leading to a successful implementation, as it bridges the gap between the vision of integration and the reality of personal adjustment.

Changing Behavior

Reflecting on these transformations, one cannot ignore the role that human behavior plays in reshaping technology. Changing the way people work is often the most challenging part of any ERP initiative.

I have observed that the success of these systems depends heavily on clear communication, comprehensive training and the fairest measure of patience. When employees are given the proper support and time to acclimate to new processes, resistance gradually wanes.

In one detailed scenario, a manufacturing client experienced significant pushback from veteran employees who viewed the change as an unnecessary disruption. Through continuous dialogue and targeted training sessions those same employees came to see the benefits of an integrated system that simplified their tasks and reduced redundancies.

The transformation was gradual, but the impact was undeniable. This demonstrates that an ERP system's true promise lies in harmonizing technology with the daily lives of the people who use it.

Overcoming Hurdles and Setbacks

An ideal scenario is hard-won. Many organizations face significant hurdles along the way, from technical setbacks to internal resistance to change. The conversation with skeptical middle managers, the meticulous customization to match specific workflows and the long hours spent adjusting the data mapping are all parts of the same process. The real accomplishment lies not in installing a system, but in re-engineering the overall approach to doing business.

As I have learned through practical experience, the integration of technology with process and people is the cornerstone of creating an ERP system that truly supports organizational goals. Each challenge that arose—from fragmented operations to the friction between standard processes and needed custom nuances—served as fuel for refining the approach. The persistent theme throughout these challenges has been the importance of building a strategy that is both realistic and forward-looking.

One memorable example involves a company that suffered multiple failed attempts at implementation. Past missteps had left a trail of disappointment, making the leadership understandably cautious. When the new ERP solution was proposed, it was met with a mix of skepticism and guarded hope. I spent hours meeting with skeptical department heads and listening to their concerns.

What emerged from those talks was a mutual understanding that no system could work miracles. The implementation required significant organizational adjustments and a willingness to adopt new practices. Today, that same organization stands stronger, operating with clarity that has generated measurable improvements in process coordination and overall performance—a testament to the importance of aligning technology with practical business needs.

Iterative Improvements

There is great merit in revisiting the concept of gradual, iterative improvement. A single, sweeping change rarely goes off without a hitch. Instead, the process is one of continuous refinement, where each phase of implementation uncovers additional insights that inform subsequent adjustments.

For instance, consider an organization that initially implemented a basic system to overhaul its production scheduling. Once this first layer was operational, the leadership saw the potential to extend the system further, to integrate inventory control, quality management and even employee scheduling. Rather than a one-time transformation, it became an ongoing process of fine-tuning that gradually built a comprehensive overview of the entire operation.

This incremental approach minimizes the risk of overwhelming the existing structure and reinforces the concept that technology should evolve in step with business growth.

Continuous Refinement

As you consider the progression of ERP systems, remember that the story is far from over. The continuous refinement and adaptation of these tools reflect the dynamic nature of business itself, where continuous improvement is not just an option but a necessity.

The transformation witnessed over the years is not just historical; it is an ongoing process that demands vigilance, pragmatic thinking and a readiness to adjust as environments shift. The insights shared here are given in the spirit of collaboration and practical guidance. This is a reminder that the challenges we face today are steps on the path toward the breakthroughs of tomorrow.

Early successes and failures in ERP implementation gave rise to a rich repository of knowledge that helped form current practices. Adopting an ERP system is not a one-off decision, but the beginning of a continuous process of refinement. Every implementation brings with it an opportunity to reassess, realign and refocus efforts in ways that drive genuine business value.

It is this continuous refinement —a persistent willingness to test assumptions and adjust plans—that makes ERP projects successful. Much like a seasoned craftsperson who refines their technique over decades, leaders in this field must be prepared to revisit and revise their strategies repeatedly to keep pace with evolving business needs.

Remember that each phase of transformation gives a chance to refine your approach. The challenges, as arduous as they may seem during implementation, may reveal the path to greater operational clarity and improved business performance.

Optimize Operations for Growth

Throughout my engagement with various organizations, I have found that the accurate measure of an ERP system's worth lies in its ability to address the intricate needs of modern business. It is not only about consolidating data or automating routine tasks; it is about optimizing operations so they can sustain growth in increasingly complex commercial landscapes.

The evolution of these systems serves as a roadmap for leaders—an ever-evolving response to the relentless pace of change. Each step forward in technology has required a parallel shift in thinking, urging business leaders to question long-held assumptions and to refocus on what truly matters: the combined power of technology, strategy and people.

A Culture of Informed Decision Making

Another key takeaway is that ERP's benefits extend beyond integrated process flow and systems performance. They contribute to a culture where informed decision-making and transparent communication become central values. When employees begin to trust the data presented to them, they are more inclined to take ownership of their roles and responsibilities.

I recall instances where even the most reluctant staff members grew to appreciate the system, not because it replaced their expertise, but because it showed them a clearer view of the impact of their contributions.

It is this kind of culture change—a shift from isolated departments to a cohesive team working toward shared objectives—that can set a company apart. The significance of this cultural transformation cannot be underestimated. It demystifies technology and makes the overall advancement toward operational excellence a collective one.

Address Issues Head On

The most successful implementations I have encountered were not the result of blind faith in a new platform but came about when leaders took a proactive approach to address important issues head-on.

Whether it was by confronting budgetary challenges that threatened to derail the project or by addressing the subtle but pervasive resistance from long-standing employees, decisive action was the common denominator.

There is a certain satisfaction in knowing that even the most complex problems can be managed with clear communication, sensible planning and a willingness to question established routines. Every time an organization manages to integrate disparate systems into a unified process, it reinforces the idea that technology, when paired with the right mindset, is a tool to serve strategic goals, not a barrier to them.

Accountability Begins at Every Level

A recurring insight throughout is that successful ERP implementation demands accountability at every level. This is not just about holding vendors responsible for technical glitches. It is about ensuring that every individual within the organization understands their role in the transformation process.

Transparent tracking of key performance metrics, coupled with regular feedback loops, allows leaders to diagnose issues swiftly and implement corrective measures before minor problems balloon into major setbacks.

The discipline of objective monitoring also reveals patterns over time—patterns that can drive long-term improvements. Realizing that even minor misalignments can have wide-reaching effects encourages decision-makers to adopt a proactive stance, constantly seeking areas where the system can be refined to better support the business aims.

In my experience, there is no shortage of examples of projects that have turned around due to determined problem-solving and rigorous, objective re-evaluation.

Projects Can Be Turned Around

In one case, a seemingly sharp divergence between projected outcomes and actual performance led the leadership to reorganize their approach completely. Rather than proceeding down a path that would have led to escalating costs and mounting frustration, the team held tough, candid discussions that eventually realigned project goals with realistic expectations. The result was a system that delivered greater functionality than originally anticipated and did so in a financially-sustainable manner.

These instances illustrate those setbacks. When met with honest reflection and a readiness to adjust the course, they paved the way for even greater achievements. Each challenge, which was overcome, added depth to the organization's understanding of its processes, which informed future decisions. Clear dialogue, consistent communication and an unwavering commitment to align every element of the system with the business's strategic objectives serve as the bedrock of these turnarounds. When a company commits to this comprehensive approach, it enables itself to cope with the complexities of the market and to seize opportunities that arise.

Insights for Executives

For the forward-thinking executive, the insights here serve a dual purpose. They provide a history of the evolution of integrated systems and a set of guiding principles for managing future challenges. Recognizing that every ERP implementation usually has custom requirements, the focus must remain on creating a system that supports the particular needs of the organization while remaining flexible enough to adapt to unforeseen circumstances.

ERP systems are as much about strategic planning and internal alignment as technology. The lessons learned—from the need for centralized data to the benefits of a modular, future-proof design—create valuable guidelines for managing and sustaining change.

A fundamental message emerging from this discussion is that the strength of an ERP system is measured not solely by its technical attributes, but by the holistic integration it facilitates. When every department and process is linked by a common set of data and shared objectives, the entire organization benefits from a clearer, more reliable operational framework.

The narrative of ERP transformation encompasses early struggles with fragmented systems, gradual progress through iterative improvements and the eventual convergence of technology, process and people into a unified, strategic asset. This progression is a roadmap for business leaders who must continuously balance the inherent risks and rewards of technological integration.

Looking ahead, it is encouraging to note that advancements in technology continue to open new avenues for optimizing business operations. However, the underlying themes that drive successful ERP implementations remain constant. The measure of success will always be the extent to which technology serves as a tool for achieving strategic objectives rather than as an isolated project with a finite lifespan. This enduring truth is the source of both inspiration and practical advice for leaders seeking to prepare their organizations for the challenges of tomorrow.

This holistic understanding of ERP systems—the centralized data, the thoughtful integration of processes, the user-centric design and the strategic alignment with overall business goals—is what transforms an implementation from a technical project into a lasting asset. It is a perspective built on decades of observing successes, learning from failures and continuously adapting to the evolving demands of the business landscape.

The narratives shared here, drawn from years of experience, emphasize that there is no substitute for a tailored strategy that respects the unique complexities of your organization. Instead, the key lies in creating a framework that evolves alongside your business, providing the transparency and strategic insights needed to stay ahead in an unpredictable market.

If there is one takeaway to hold onto, it is that the true power of an ERP system is realized when technology, practices and people align. The stories and lessons recounted throughout this exploration serve as a reminder that while the technical components are a key factor, the human element remains at the heart of any successful initiative.

Fostering an environment where every employee understands their role and has access to reliable information sets the stage for a more resilient, agile and successful organization.

By applying the principles of centralized data, iterative improvement and a thorough understanding of both technological and human factors, you can craft a strategy that propels your organization toward lasting success.

As you move forward, let the insights shared here serve as both a foundation and a compass. The lessons learned, from early missteps to triumphant turnarounds, are a testament to the fact that with careful planning, honest reflection and the dedication to align every aspect of your operations, a more coherent and effective business structure is well within reach. In that alignment lies the promise of not just surviving challenges but thriving amid the uncertainties of tomorrow.

The implementation process, with its twists and turns, reaffirms that every decision made in pursuit of clarity and efficiency builds toward a stronger, more adaptive organization. Understanding that the path forward is paved with continuous learning and steadfast execution, business leaders can face the future with confidence, equipped with the insights necessary to turn each challenge into a successful step toward operational mastery.

CHAPTER 2

THE CASE FOR ERP IMPLEMENTATION

Chapter 2: The Case for ERP Implementation

ERP systems evolved from isolated tools into unifying platforms capable of transforming business operations. With that understanding as our foundation, the next step is to consider why organizations must adopt these systems in the first place.

In a competitive environment, every decision carries weight, and the decision to invest in an ERP solution is no exception. The challenge lies in recognizing the symptoms of operational inefficiency and then identifying a solution as a part of the overall business strategy.

The Need for ERP

Business leaders face a critical juncture where they must assess current performance, question long-standing practices and unequivocally decide that the path forward demands a new framework. This calls for an honest examination of what works, what doesn't and what changes are necessary to secure a lasting success.

I recall my first encounter with an ERP project for a manufacturing company. The organization was starving for information that hindered customer service and disrupted internal coordination. Different departments operated on separate timelines, and data was locked away in silos, rendering the executive team nearly blind to the workings of their business. This was a wake-up call.

Management understood that a coordinated effort was required to stitch together every facet of their operation into a cohesive whole. Despite the skepticism and the overwhelming fear of untested changes, the ERP project moved forward. That project was a learning experience illustrating the promise of integrated systems and the potential pitfalls if the broader implications were not recognized.

Recognize the Symptoms

Identifying the need for an ERP system starts with an honest look at the current state of operations. From fragmented data sets and communication breakdowns to the mounting costs of inefficiency, the signs are all there—clear signals that the status quo is no longer sufficient.

Recognizing these symptoms is the first step in a broader process of transformation, one that aims to align technology, processes and people into strategic collaboration.

The insights shared here are meant to serve as a guide as you begin your own assessment. They remind you that every inefficiency, every miscommunication and every delay is not just a problem, but rather a window of opportunity. The decision to invest in a comprehensive, integrated system is a strategic imperative that can unlock unprecedented value for your organization.

As you move forward, carry these insights with you. Use them for discussions with your teams and let them be the foundation upon which you build a compelling case for change. In doing so, you address the symptoms observed today and set the stage for a future where every aspect of your organization works in unison to drive growth, clarity and sustained success.

In many ways, recognizing the need for change often starts with listening to the subtle signals emerging from within an organization. There comes a point when operational delays, mounting frustrations and persistent inefficiencies can no longer be ignored.

Over my years of working with diverse companies, I have noticed these signals emerging in a variety of ways. They may come as misaligned data that throws off executive decisions, or as persistent disconnects between departments that lead to frustration among employees. These early indicators are not just isolated issues, but symptoms that the current systems are not fully supporting the strategic goals of the business. Addressing them requires leaders to look beyond temporary fixes and toward comprehensive solutions that will enable sustained growth.

I recall a medium-sized enterprise grappling with repeated issues that threatened both its efficiency and its bottom line. Data was fragmented, and reporting was delayed by an entire business cycle, creating a sense of unease among senior management. The fragmentation wasn't due to a lack of effort; rather, it was the cumulative result of operating with a series of disconnected systems built over time.

Each department had developed its own way to capture and process information, resulting in a hodgepodge of data that made it nearly impossible to obtain a clear, real-time view of the organization's performance. That situation became a turning point, a moment when decision-makers recognized that a more integrated approach was not simply desirable, but vital to maintain competitiveness and operational clarity.

Begin with an Assessment

Leaders must take an honest look at their current operational processes and determine whether these processes are truly serving the company's strategic goals.

This methodical evaluation often reveals that many persistent symptoms—such as repeated data inaccuracies or delays in order processing—stem from the inherent limitations of the existing processes and systems supporting those. These issues are not solved with patches or temporary workarounds; they require a fundamental reassessment of how information flows within the organization.

One practical approach I have encouraged is to map out the entire process landscape, visualizing how data and operations interact across the organization. When you can see how a delay in one department creates problems in another, the case for a unified approach becomes much more tangible.

In one instance, a company struggling with inventory management realized that the root cause of frequent stockouts was not simply poor planning, but a series of communication gaps across its supply chain. Once the leadership mapped out each step of the process, the interconnected nature of the delay became clear. This newfound clarity underscored the need for an integrated framework that would create a single source of truth and enable better coordination between departments.

Document Challenges

In practical terms, diagnosing the need for a comprehensive solution begins by engaging all levels of the organization in a candid conversation and documenting current challenges. This conversation should be grounded in factual data as well as observations. When leaders, managers and front-line employees contribute their perspectives, a more complete view of the operational landscape emerges.

One of the most effective ways to identify these challenges is through structured reviews and process audits that solicit input across the organizational hierarchy. These exercises produce concrete evidence of inefficiencies and help to build the internal consensus needed to support a major transformation.

I remember an engagement where I led a series of workshops across various departments of an organization that had been suffering from disjointed processes. It was during these sessions that the full extent of the operational challenges became apparent. Employees, who had long felt that their voices were not being heard, shared stories of repeated delays and frustrating miscommunications.

Once these symptoms were unpacked and documented, the leadership team was able to quantify the cost of these inefficiencies and build a persuasive case for change. The result was an unwavering commitment to invest in a fully-integrated system that addressed the root cause rather than just the symptoms.

The process of diagnosing these operational challenges is, therefore, not just a technical exercise; it is a strategic one that lays the groundwork for a broader transformation. It forces leaders to confront hard truths about the current state of operations and to recognize that incremental improvements are not enough when the objective is long-term sustainability and growth.

By taking a holistic view of the organization's processes, you can identify areas where the cost of inaction far outweighs the investment needed for a modern ERP system.

Replacing Entrenched Systems

I have witnessed how entrenched systems, which once performed admirably, eventually become liabilities as the business scales. When performance metrics start to lag behind forecasted goals and when internal feedback consistently points to miscommunication between departments, the need for a more cohesive approach becomes undeniable.

A telling example involves a company where each department was using its own metrics for tracking performance, which led to a fragmented understanding of the business's overall health. Without a unified perspective, top management found it difficult to pinpoint the source of delays or inefficiencies.

This experience reinforced an essential lesson—when every part of an organization clings to its own legacy processes without a broader, strategic alignment, the collective performance suffers. It becomes increasingly challenging to implement changes that would benefit the organization as a whole. In such cases, the step toward a comprehensive solution is not simply about adding another piece of technology; it is a strategic pivot that reshapes how every department communicates and collaborates.

Transformative Change

It is often during challenging times that the need for a unified system crystallizes. I have witnessed organizations teetering on the edge of frustration, with fragmented data, misaligned processes and siloed departments impeding timely decisions.

In one memorable encounter, a company struggling to manage its financial processes was caught in a perpetual loop of delayed reports and reactive decision-making. The chaos created fertile ground for missed opportunities as each segment of the business operated with its own agenda, often out of sync with the actual situation.

The leadership recognized that filling the gaps required more than incremental fixes—it required a fundamental overhaul of existing systems. This realization was a turning point; an acknowledgment that the way forward lay in creating a centralized system capable of delivering clarity and unified insights.

Strategic Redesign

Understanding the need for ERP goes beyond simply noting operational challenges. It involves a deep dive into the underlying causes of those challenges. Many organizations wrestle with processes that have evolved organically over time—incremental adjustments made to address immediate obstacles rather than strategic redesign.

In these cases, the symptoms are evident, which include outdated methods for tracking inventory, recurring data errors in financial reporting and communication breakdowns between departments. While these issues are often dismissed as inevitable nuisances, they can erode a business's competitiveness over time.

The case for ERP implementation is built on the recognition that addressing these challenges through an integrated approach remedies current pain points and positions the organization for scalable and sustainable growth.

I have seen how a misalignment between business strategy and the tools in place can lead to stagnation. Consider a company that, despite a strong vision for growth, continued to rely on legacy systems that no longer supported its aspirations. The sheer complexity of reconciling multiple databases and incompatible software applications slowed down processes and generated an air of uncertainty among employees and stakeholders.

In such scenarios, the benefits of a unified ERP system become evident. A well-integrated platform can streamline processes, establish a clear link between data and decision-making, and, most importantly, support a culture of proactive management.

For any executive, the fundamental question becomes: How can technology be harnessed to serve the broader vision and drive the organization forward? This is the essence of building a business case for ERP investment—a process that entails a rigorous evaluation of both risks and rewards.

Shift in Organizational Mindset

It is essential to recognize that the decision to implement an ERP system is often accompanied by a significant shift in the organization's mindset. The path toward a unified system invariably involves confronting entrenched habits and confronting the perception that "this is how we have always done it," which impedes improvements.

In my experience, the leaders who successfully pivot are those who can articulate the strategic vision in a way that resonates with every layer of the organization. They communicate that the objective is not to discard past efforts but to build upon them in a manner that captures the full potential of the business.

This cultural shift is as important as the technological upgrade itself. The true test of leadership lies in motivating the entire organization to view the ERP project not as a disruptive force, but as a necessary evolution—a chance to realign everyday operations with long-term goals.

The Gradual Erosion of Efficiency

This introspection is critical because the symptoms are often subtle and build up over time until they lead to significant inefficiencies. Fragmented systems might not cause dramatic failures overnight, but they steadily erode the potential efficiency of the entire operation.

It is the accumulation of these minor setbacks—lost hours reconciling differences between spreadsheets, missed opportunities due to delayed reports and constant firefighting that results from data discrepancies—that motivates leaders to take a decisive step towards integration. When every day adds up to a few more wasted resources and mounting frustration, that is when the true cost of inaction becomes apparent.

Bridge Between Strategy and Execution

The process of identifying the need for a comprehensive solution is, at its core, an exercise in aligning execution with strategy. The symptoms that accumulate over time—data silos, recurring delays or miscommunications—are the early warning signs of a system under strain from outdated practices. Recognizing these signals and translating them into actionable insights paves the way for the transformation to a new ERP system.

In many ways, this process is a microcosm of broader strategic planning, where the objective is not simply to react to problems, but to reimagine how the organization can work more effectively in the future.

When I look back at my engagements with organizations struggling to manage change, it becomes clear that bridging the gap between strategy and operations is a critical piece of the puzzle. Over time, I noticed that while symptoms of disjointed processes are readily apparent, the real challenge lies in aligning strategic objectives with the underlying systems that drive daily operations.

Aligning ERP Goals with Business Strategy

Remedying isolated inefficiencies is not enough; the goal must be to build a framework that ushers in a renewed sense of purpose and unity across all functions. Let's explore this process further, unraveling how carefully aligning technology with long-term strategic goals not only remedies immediate issues but also sets the stage for sustainable growth.

A recurring theme I encountered is that many organizations operate in isolation from their strategic vision. Frequently, departments remain so focused on their own operational challenges that they lose sight of the overall business goals.

Consider one organization where the sales team operated using separate tools, while finance used legacy methods passed down over many years. This created an environment with fragmented data and where decisions were made in silos. Leaders realized that without a centralized system to harmonize data, the company's strategic vision was undermined.

In that instance, the convergence of disparate systems into one platform created a single source of truth. The transformation was both immediate and profound: decision-makers were able to view insights from every angle, and everyone started to work toward the same strategic goals.

Strategic Alignment

The discussion about the case for ERP implementation naturally leads to the topic of strategic alignment.

The promise of an ERP system is not achieved by installing new ERP software; instead, success hinges on how well that ERP software aligns with your company's strategic objectives.

Every organization has its own set of challenges, and a one-size-fits-all approach seldom effectively meets those needs. Tailoring the system to fit your specific operational landscape involves deep analysis, honest discussions with key stakeholders and a willingness to reengineer existing processes. Alignment between technology and strategic objectives is not a one-off achievement but a sustained process with profound implications for every aspect of the business. It transforms isolated departments into interconnected teams with streamlined processes and creates a foundation for growth that is both resilient and adaptable.

Achievement of the goal requires time, investment and a willingness to confront ingrained habits head-on. Yet, the benefits extend far beyond the initial implementation. With a unified system in place, organizations experience greater operational efficiency and enhance their ability to innovate while responding to shifting market conditions.

As business environments grow more complex and competitive, the imperative to align technology with strategy becomes even more pressing. Waiting for isolated problems to escalate is a risk that no forward-thinking leader can afford to take. Instead, by investing in an integrated ERP system that is carefully aligned with long-term goals, companies can secure a competitive edge and be well-equipped to handle both current challenges and future uncertainties.

The Diagnostic Phase

The diagnostic phase establishes a shared understanding of the problems at hand among key stakeholders. When executives, managers and staff come together to discuss what is not working, they confirm the need for change and form a collective vision for what success might look like.

The diagnostic phase also acts as a precursor to broader strategic planning. Once the symptoms are clearly identified and understood, the next step is to articulate what success will look like. What improvements in efficiency, responsiveness and decision-making will a unified system bring?

With clear criteria in place, the case for investment becomes anchored in measurable improvements rather than abstract promises of future success. Leaders can then align their teams around these shared objectives, understanding that technology is just one part of a larger puzzle that includes redesigning processes, employee training and cultural change.

As you reflect on your own operations, keep in mind that the goal is to create a unified operational framework that supports current activities and the long-term vision of the company. The challenge is to move beyond temporary workarounds and to build a system that is resilient, flexible and aligned with strategic priorities. This means that every decision you make—from reengineering a process to investing in new technology—should be viewed through the lens of how it contributes to a larger, integrated approach.

The symptoms you observe are not just problems to be fixed; they are opportunities to reframe the narrative of your organization's operations. They create a chance to address longstanding inefficiencies and to implement changes that will resonate throughout the business, reinforcing every department's ability to act decisively and collaboratively.

It is a call for analytical and humane leadership, acknowledgment of what isn't working and investment in a future that operates as one cohesive unit, where every part of the organization is in tune with the same strategic objectives.

The Value of Operating with Real-Time Data

An improved environment is where every piece of information—from customer orders to inventory levels—is captured in a central repository and accessible to decision-makers as self-service. Such a setup is far more than a technical upgrade. It represents a fundamental shift in the way business is conducted. This approach can illuminate weak spots in current processes and reveal opportunities for streamlining operations that might otherwise remain hidden.

From a leadership perspective, the case for implementing a unified system is compelling when it comes to achieving sustainable growth. The decision is not just about the technical merits of one software solution over another. It is about whether the existing operational model can support strategic ambitions. If you find that every attempt to improve specific areas results in new challenges cropping up elsewhere, it is a signal that the underlying framework is misaligned with the demands of the modern marketplace.

Articulate What Success Looks Like

At its core, achieving strategic business alignment requires a candid conversation between technology leaders and business leaders—a conversation that goes beyond technical specifications to consider the organization's business ambitions. One of the first steps in this process is to articulate what success looks like in concrete terms.

Being able to articulate concrete benchmarks for success—such as reducing order-processing times, improving forecasting accuracy or increasing overall operational responsiveness—creates a foundation upon which the business case for ERP becomes unassailable. It transforms what might have been a technical upgrade into a strategic initiative with measurable benefits. These clear metrics help everyone see that the investment is not a cost in isolation, but an engine for future growth that will realign every department with the company's overarching objectives.

For instance, leaders might define success as reducing the work-in-progress inventory by a certain percentage, or as achieving a measurable improvement in the cash-to-cash cycle. These benchmarks create a clear target and serve to anchor expectations throughout the implementation process. I have seen leaders who initially approached ERP as a purely technical upgrade shift their perspective once they discussed these strategic performance indicators. In doing so, they were better positioned to evaluate the long-term impact of their investments.

To illustrate, I recall working with an organization that struggled with forecasting demand for its products. The existing system produced reports that were often outdated, making it difficult to adjust to sudden market changes. Once the leadership examined the issue through a strategic lens, it became clear that the underlying problem was not solely technical and procedural. The reports were generated too slowly, and their structure did not facilitate timely decision-making.

By redefining what forecasting success meant for them—speed, accuracy and actionable insights—the organization was then able to pinpoint exactly where the process faltered and how a unified system could bridge that gap. The eventual outcome was a system that delivered near real-time forecasts, directly impacting inventory management and customer satisfaction. It was in moments like these that the benefits of a strategic alignment became apparent.

Methodical Review of Existing Processes

Bridging the gap between a company's vision and its operations also demands a methodical review of existing processes. This review is more than a technical audit. It is a strategic exercise that digs into how each business unit contributes to the overall vision.

When I worked with a multinational firm, one of the initial steps was to create detailed process maps for each function. These maps laid bare inefficiencies and highlighted areas where miscommunication was causing delays.

As an example, the production team failed to synchronize its schedule with sales forecasts, causing either overproduction or stockouts. By visualizing these operations across functions, it became apparent that the current framework was a patchwork of ad-hoc solutions rather than a cohesive system. The resulting discussions led to a reimagined process that was built into the new ERP system, one that recognized that each department's workflow was supported and enhanced the overarching strategy.

Such exercises also underscore the importance of setting realistic expectations regarding the transformation. Change, as I have learned from years of experience, tends to lead to temporary disruptions. There are early challenges—such as resistance from staff who feel their accustomed routines are under threat—or the inevitable bumps that occur with new system rollouts. However, the goal is to view these initial setbacks as integral parts of a longer process rather than as failures.

In one instance, a company experienced significant delays at the outset because the integration of its legacy systems with the new software was more complex than anticipated. Instead of abandoning the project or resorting to quick fixes, leaders chose to address the root causes by investing additional time in system testing and employee training.

Gradually, these investments paid off, and the system began to align more closely with strategic objectives. The key was not to shy away from the challenges but to confront them head-on with a clear-eyed focus on long-term goals.

ERP Systems as a Change Agent

Another dimension worth exploring is how ERP systems can act as a catalyst for broader cultural change within an organization. When a new system is introduced, it is not just a shift in technology and an opportunity to reset how teams think about their work.

Employees may often be entrenched in what they perceive as "the way it has always been done." Breaking through that mindset requires empathetic and decisive leadership.

In one project I led, the implementation of a new ERP system was met with skepticism by long-standing employees who feared that the system would add layers of bureaucracy or threaten their established roles. Through a series of open forums and hands-on training sessions, leadership communicated that the change was designed to enable them rather than constrain them.

Over time, as employees began to see the tangible benefits—such as faster information flow and reduced manual errors—the cultural shift took hold. The new system became a tool for collaboration rather than a point of contention. This change in perspective is invaluable because it transforms a technical upgrade into a transformative reorientation of the entire organization.

Champion of Change

Leaders must be prepared to champion change, ensuring that employees at every level understand the broader vision and the benefits of moving to an ERP Solution.

As you examine the symptoms within your own organization, consider how persistent challenges might be indicators of systemic inefficiencies. The signs are often interwoven with broader issues of accountability, communication and process alignment. When employees report that they are working in silos, or when efforts to improve one part of the system inadvertently disrupt another, there are clear signals that the current framework is failing to support the organization's goals comprehensively.

It is in these moments that the rationale for a more integrated approach becomes not just compelling, but essential.

External Stakeholders

In addition to internal culture, aligning technology with strategy also means rethinking external relationships. The contemporary business environment is dynamic, and external partners—be they suppliers, customers or service providers—play an integral role in an organization's success. Modern ERP systems can extend beyond internal operations to create platforms for wider collaboration.

By sharing real-time data with suppliers, for example, companies can improve supply chain coordination and respond more adaptively to market demands.

I recall assisting a company that had long dealt with inconsistencies in its supply chain. By integrating its ERP system with key supplier systems, the company achieved a level of transparency that dramatically reduced lead times and error rates. This external alignment enhanced operational efficiency and strengthened partnerships that were critical to the overall strategy.

This process of external alignment is sustained by the recognition that a well-integrated system is not an endpoint but part of an ongoing evolution. As markets change and new competitors emerge, organizations must be nimble and responsive.

The Transformation

The case for ERP goes far beyond addressing operational inefficiencies—it is a strategic initiative designed to fundamentally reshape how an organization functions. By meticulously aligning technology with the broader vision of the company, leaders set the stage for transformation that touches every corner of the business.

This alignment is a powerful tool, one that has the potential to convert latent inefficiencies into growth engines, turning everyday challenges into opportunities for improvement and innovation.

The key takeaway is that the decision to realign technology with strategic objectives is not just a technical upgrade. It is a deliberate reimagining of how information flows, how decisions are made and how every department contributes to the company's long-term success.

Organizations that take this approach are better positioned to harness the full capabilities of modern ERP systems by transforming them from collections of data and modules into cohesive, strategic assets that drive consistent performance improvements.

Continued Improvement

The implementation of an ERP system is not a once-and-done initiative. It is the foundation upon which continual refinement and adaptation are built. Leaders who embrace this view recognize that periodic reassessment of processes and updates of system configurations are part of the process.

In one dynamic market I observed, a company that initially focused solely on internal alignment later expanded its ERP integration to include customer relationship management and external vendor interactions. This streamlined operations and created a more holistic view of the business ecosystem, ensuring that every decision was substantiated by comprehensive, up-to-date data.

One key lesson I learned from various projects is that the conversation does not end once the system is in place. It begins anew with every cycle of feedback and analysis. Metrics improve incrementally, resistance fades as wins accumulate and the organization starts to enjoy the rewards of decisions made based on reliable, unified data.

This continuous improvement culture is integral to ensuring that the new system remains more than a static tool—it becomes a living framework that adapts to evolving market conditions and internal organizational shifts.

This enduring commitment to iterative learning helps transform the entire business's operations. When leaders have built in processes for regular system reviews, they are better positioned to identify emerging bottlenecks before they have a chance to escalate.

In one case, a company that had once struggled with periodic system slowdowns began using its own data to predict underlying trends in operational delays. This proactive use of technology preempted potential issues and contributed to operational innovations that propelled the business forward. In that environment, every challenge became an opportunity to refine processes further, ensuring that short-term setbacks were converted into long-term strategic advantages.

Building a Business Case for ERP Investment

Through clear-eyed analysis and strategic foresight, building the case for ERP becomes a transformative exercise—one that bridges the gap between business strategy and execution. It encourages a proactive stance, setting the stage for a future where every decision is supported by reliable, up-to-date information.

The objective is to turn uncertainty into opportunity, using measurable insights and deliberate planning as the cornerstones of a well-informed investment.

With this understanding, the path forward reveals itself not as a leap into the unknown but as a series of deliberate, well-curated steps. Each step invites scrutiny of existing processes and calls for honest appraisal of where improvements can be made. The result is a compelling narrative that justifies investment and inspires confidence among all stakeholders.

This moment is a call-to-action for leaders determined to move beyond reactive measures. It is a juncture where the data, the insights and the hard-won lessons from past implementations converge to form a clear view: the time to act is now. By confronting the realities of operational inefficiency head-on and building a sound business case for ERP, you place your organization in a position to address current shortcomings and to seize emerging opportunities with confidence.

Throughout our discussions, a clear thread has consistently emerged as the decision to invest in an ERP system is not simply about installing new software or making incremental fixes. It is about setting the organization on a path where every process, decision and relationship is anchored in a cohesive framework that directly supports its long-term strategic objectives.

Persistent symptoms, which include fragmented data flows, siloed processes and misaligned departmental goals, serve as both a warning and an opportunity. They signal that the business, as it stands, is being held back by outdated practices and systems that have become more of a burden than a benefit.

The Cost of Misalignment

Throughout my career, I have observed numerous organizations lose valuable time, resources and competitive edge simply because they failed to recognize that these symptoms, however minor in isolation, accumulate over time to thwart a company's broader ambitions. In these experiences, the value of a unified system is revealed not in its ability to fix individual problems but in its transformative power to turn every function of an organization into a well-coordinated, strategic contributor.

The early signs of misalignment—chronic delays, recurring data inaccuracies and miscommunications—were not isolated glitches. Instead, they pointed to a deeper problem. The existing system was pieced together through decades of ad-hoc responses rather than thoughtful design.

Instead of being a competitive strength, the disjointed nature of these processes stifled growth, creating a cycle of inefficiency that caused tangible costs to the bottom line and reduced the morale of the teams involved.

I have seen how an organization, when faced with the mounting consequences of these challenges, can undergo a significant transformation by stepping back and reassessing its operational framework. The first step involves careful diagnostics—mapping out every process, scrutinizing both internal workflows and external partnerships and identifying the exact points at which inefficiencies seep in.

In one notable case, a company that once struggled with forecasting and inventory management managed to reveal that its issues were not simply technical errors, but the result of a cascading failure in communication between sales, production and supply chain teams. This realization was not just a call for a technical upgrade but a demand for an entirely new approach, which was creating a proactive and unified system that turned fragmented insights into a clear, actionable strategy.

Benefits of an Integrated System

A unified system, by its nature, supports better integration with external entities. Modern ERP solutions are designed to connect with suppliers, partners and customers. By extending the flow of accurate, timely information outside the organization, companies can enhance supply chain efficiency, drive better customer relationships and create stronger vendor partnerships.

I recall engaging with a client who had long contended with inconsistent supplier performance, a consequence of mismatched data and disjointed workflows. Once their ERP system connected them directly with key suppliers, discrepancies in orders and deliveries diminished. This integration streamlined operations within the company and reshaped its external relationships, demonstrating that the benefits of alignment have far-reaching implications.

It is not uncommon for organizations to struggle with the uncertainties of rapid change. In business, where speed and adaptability are paramount, traditional manual processes or isolated systems simply cannot keep pace. The symptoms that eventually prompt a search for better solutions often include chronic issues such as recurring delays in order processing, inconsistent data across different business units and the frustration that comes from constantly having to reconcile conflicting information.

These recurring problems serve as a reminder that the operational model in place was designed for an earlier era, one when businesses were smaller and less interconnected. What once worked efficiently may now be the very obstacle to progress. There are numerous examples where companies rested on their laurels, assuming that their processes and market positioning were top-notch. Barely a few years later, such companies came crashing down like a house of cards as their revenue shrank rapidly in line with their affinity to stay in their comfort zone.

In many cases, the decision to invest in an integrated solution is triggered by business needs, but it is more than just a reaction to immediate pressure. It is driven by a strategic recognition that consistent, real-time data is essential to track performance and to anticipate changes in the market.

Risk vs. Reward

As you consider this phase of exploring the need for ERP within your organization, it is helpful to reflect on the balance between risk and reward. On one hand, there is the potential for operational disruption during the transition phase—a period marked by adjustments, temporary slowdowns and the inevitable learning curves. On the other hand, the benefits, once realized, can be transformational and usher in improved coordination, better decision-making and enhanced agility in response to market dynamics. It is through a rigorous analysis of these factors that business leaders can make the most-informed decision.

The objective is to chart a course that minimizes risk while maximizing the strategic and financial benefits of a unified operational framework.

The opportunity here is not just to upgrade a system, but to fundamentally recalibrate how decisions are made at every level of the organization. Recognizing the full spectrum of benefits—from accountability and transparency to agility and market responsiveness—creates the impetus to invest in a solution that aligns tightly with business strategy. In doing so, a business does more than consolidate data; it crafts a narrative for its future where every operational unit works in concert and each decision is underpinned by precise, real-time insights.

Short-Term Fix vs. Strategic Investment

At this juncture, it is essential to differentiate between short-term fixes and long-term strategic investments. Short-term solutions often involve temporary adjustments or ad-hoc improvements that address immediate concerns but fail to eliminate the underlying issues. Over time, these piecemeal efforts can lead to a patchwork of solutions that only serve to complicate the operational framework further.

By contrast, a well-integrated system lays the foundation for a comprehensive, long-lasting improvement. I encountered situations where companies, desperate for immediate relief, opted for quick fixes that only delayed the inevitable need for a broader system overhaul. While these stopgap measures might create a momentary respite, they rarely solve the deeper challenges embedded within the organization.

The long-term view, however, requires a willingness to invest in a solution that is aligned with the company's future ambitions. This involves rethinking existing processes along with implementing the technology. It is essential to set expectations so that every component of the organization is prepared for a unified way of operating.

In this context, an integrated solution is not an end in itself; rather, it is a critical enabler that sets the stage for a more agile and responsive business model. The challenge then becomes one of ensuring that investment is made in technology and in the cultural shift needed to support it.

Return on Investment

One cannot ignore the financial implications of these decisions. The numbers don't lie, and in many cases, the return on investment from a well-executed ERP implementation can be substantial. However, making that case requires more than a spreadsheet filled with projections; it requires a deep understanding of where inefficiencies lie and how they negatively impact the bottom line.

I recall a scenario where a CFO found that costs were spiraling due to duplicated efforts across departments, all masked by outdated reporting tools. When the opportunity to consolidate and streamline these processes was presented, the potential savings were striking. That realization, however, was only possible through a detailed analysis that connected operational inefficiencies with financial losses. It underscores the reality that every dollar lost due to ineffective processes is a signal to reassess the status quo.

A well-constructed business case for ERP should combine these financial insights with the tangible benefits of improved operational performance, demonstrating that investment is not a lost expense but a strategic enabler of sustainable growth.

Thrive in a Dynamic Marketplace

An organization's capacity to thrive in a dynamic market depends on its willingness to confront emerging challenges with bold, calculated changes. The need for a unified system is not driven by technology alone, but by the combination of precise, objective data and the pressing demands of strategic growth. When every process is interconnected, every decision is supported by actionable insights and every stakeholder—internal or external—operates from a shared understanding, the organization is positioned for success.

These reflections are an invitation to view any current challenges as opportunities. The path to implementation described here is one of rigorous analysis, measured change and a lasting commitment to aligning every aspect of the organization with its strategic goals. The insights shared through these pages enable every leader with this knowledge to recognize the signs of inefficiency, to build a compelling business case for change and to drive the transformation that secures the organization's future.

Multidimensional Rationale

In synthesizing these experiences, it becomes evident that the rationale for ERP implementation is multidimensional. There is no single fix that transforms an organization overnight. Instead, there is a comprehensive reimagining of how every aspect of the business operates, from internal processes to external interactions.

The decision to invest in a unified system—and to do so with strategic clarity—is reflected in measurable improvements in operational performance, cultural transformation and overall growth potential. The narratives shared here, drawn from years of working with organizations of various sizes, echo the same message, which is, a cohesive, integrated system is not a luxury, but a fundamental necessity for survival and success in a complex business environment.

Compelling Business Case

The practical steps toward constructing a compelling business case for ERP implementation require a blend of hard facts and an intuitive understanding of your company's dynamics. Leaders must be ready to quantify the current costs of fragmented systems and link them directly to missed opportunities in the market. Often, this involves gathering data that may have been overlooked in routine operations, such as a pattern of delays, the frequency of errors or even the degree of customer dissatisfaction.

Transforming these observations into a coherent narrative that justifies the need for a new approach is as much an art as it is a science.

I have seen executive teams come together, examining detailed analyses and having long hours of discussions, to unravel the hidden costs that were silently draining organizational performance. The results of such exercises are eye-opening, turning what might have been just a technical upgrade into a comprehensive strategy for rejuvenating the company's competitive edge.

Beyond the internal dynamics, the external market environment also plays a pivotal role in shaping the decision to adopt an ERP solution. In a globalized economy, speed and accuracy in responding to market changes are not optional; they are essential. A company that hesitates to modernize its operational systems may find itself outpaced by competitors who are far more agile, with real-time data and streamlined processes supporting their strategic decisions.

I have spoken with executives who expressed regret for not embracing a more integrated system sooner. The delay cost them, in the form of missed revenue opportunities and erosion of market share to more technologically adept rivals. The imperative, then, is clear, which is, to remain competitive, the decision to invest in ERP must be underpinned by a thorough understanding of both internal challenges and external pressures.

The case for ERP is strengthened when it is seen as a critical step toward future resilience rather than just a remedy for past inefficiencies.

The question is not whether a business should invest in ERP; rather, it is how soon it must act to remain competitive. An effective ERP system is the backbone of operational excellence, empowering organizations to thrive in an evolving marketplace. Business leaders who champion ERP investment position their companies for sustained growth, resilience and digital maturity.

ERP is not a mere software solution—it is a catalyst for strategic success. The time to invest is now.

CHAPTER 3

PREPARING FOR ERP IMPLEMENTATION

CHAPTER 3: PREPARING FOR ERP IMPLEMENTATION

The previous chapters illuminate why an integrated ERP approach is not a luxury but a necessity, as a response to fragmented systems and unsynchronized processes. Building on the insights about the need for an ERP system, it is time to shift the focus from the strategic rationale and diagnosis of operational inefficiencies to the concrete steps required to put change into motion.

Build Confidence

One of the most valuable outcomes of a rigorous preparation phase is the building of internal confidence. When each stakeholder understands how their role contributes to the broader transformation, it creates a shared sense of responsibility and commitment.

In one of the projects I led, I saw how regular, strategic meetings—focused on clarifying goals, addressing concerns and refining plans—helped unify the team. These meetings improved communication and provided multiple opportunities to identify risks and develop strategies to mitigate them. The result was a highly resilient implementation plan that succeeded in meeting its objectives despite several external obstacles.

This experience shows that preparation, when done thoroughly and inclusively, can transform a technical project into a unified strategic initiative that engages everyone in the organization.

Moreover, effective preparation sets the tone for the entire implementation. It creates a structure where problems are anticipated, and responses are pre-planned, reducing the chaos that often accompanies major transitions. By establishing clear milestones and metrics to monitor progress, you create an environment where continuous improvement is not just an ideal but a practical reality.

These metrics, which can range from process efficiency and cost savings to employee satisfaction and system performance, form the compass by which you manage the complexities of implementation. They can also be used to create tangible evidence of progress, serving as checkpoints that reinforce the value of the efforts invested during the preparation phase.

Your Roadmap to Success

As you consider these insights, it is worth emphasizing that they create a strategic roadmap meant to guide leaders through the intricacies of aligning technology with a clear vision.

Every piece of evidence gathered, every process mapped out and every conversation held has been a step toward building a stronger, more resilient organization. The experiences described here are a testament to the fact that addressing inefficiencies and connecting disparate systems is a powerful catalyst for long-term success.

The benefits of this strategic alignment are best understood through the lens of transformative change. A unified system brings together data from every corner of the business and creates an environment where accountability and informed decision-making are the norm.

When all functions operate from a common set of facts, decisions become more rational, timely and aligned with the company's overarching goals. The cumulative effect of these changes is a business that is far more agile, able to anticipate market shifts and capable of responding with precision and confidence.

The aim is to discover a comprehensive perspective on these challenges and opportunities—one that integrates the more granular aspects of operational inefficiency with the broader strategic imperatives of modern business leadership.

It is essential to recognize that creating a more integrated, unified system is an ongoing process—one that requires sustained attention, continuous learning and an unwavering commitment to strategic clarity.

This chapter sets the stage for turning diagnosis into action—ensuring that every decision made today will contribute to a more coherent, agile and resilient organization tomorrow. Identifying operational inefficiencies, aligning technology with long-term goals and building a compelling business case for change is a critical prerequisite for the next stage of transformation.

With that in mind, you are now ready to transition to the practical phase of planning and preparing for ERP implementation—where the visionary insights of strategy combine with actionable steps to secure lasting success.

The Value of Extensive Preparation

Over the years, I have observed that even the most compelling business cases fail to deliver results if the preparatory phase is overlooked or underestimated.

I recall a project with a medium-sized manufacturer that, under intense market pressure, was prepared to implement an ERP system in record time. Despite high-level enthusiasm, they initially struggled because their project team was hastily assembled and lacked a clear, mutual understanding of roles and responsibilities. What eventually saved the project was the realization that the foundation of a successful ERP implementation is not a race against time, but a carefully prepared structure that aligns every team member with the overall vision for change.

This experience taught me that the strength of the implementation relies on the team's collective capability and the precise planning that occurs long before the software goes live. In practice, the preparation phase is like laying the blueprint for a complex construction project.

Before you can build a system that serves every facet of your organization, you must be certain that all elements—from personnel and processes to timelines and budgets—are in harmony. The conversation shifts from "Why do we need this change?" to "How do we get there?"

This stage is about laying the groundwork for that complex transformation. The objective here is to prepare your organization for a successful ERP implementation by focusing on assembling the right team, establishing realistic timelines and budgets, breaking the communication barriers between the team members to foster team spirit and thoroughly reviewing your business processes. These are the practical steps that turn strategic insight into actionable plans.

Reengineering Organizational Operations

The case for implementation of an ERP system becomes an invitation to reengineer your organization's operations, ensuring that every aspect of the business, from the shop floor to the executive boardroom, works in unison toward common, strategic goals.

As you prepare to evaluate your organization's readiness for transformation, consider this process an open dialogue. This process is a shared commitment to understanding that change, while challenging, is necessary for progress. Aligning strategic objectives with practical, real-world observations makes a compelling case that underscores the importance of proactive investment in technology. This approach is not about chasing trends. It is about recognizing and acting upon the ever-present need to adapt, refine and secure a competitive advantage in the face of constant change.

With a clear understanding of these dynamics, the stage is set to explore the ingredients for a successful ERP investment. By assessing current shortcomings and projecting the benefits of a cohesive, integrated system, leaders can make decisions that safeguard the future of their organizations.

Look Deeply into Processes

In my own experience, the most successful transformations have always been those where leadership did not shy away from looking deeply into the processes that underpinned their operations. Instead, by taking a hard look at where data is generated, where delays occur and how decision-making is affected, leaders have been able to create a narrative that clearly justifies a comprehensive, integrated solution. This narrative, supported by both qualitative insights and quantitative benchmarks, becomes the bedrock upon which a truly unified organization is built.

As you move forward, I encourage you to use these insights as a framework for assessing your own operations. Reflect on the gaps between your current systems and your strategic vision. Consider how the investments you make today can serve as a foundation for the resilience and adaptability needed for tomorrow. The process of aligning technology with strategy is ongoing, and every step taken—every challenge faced—is an opportunity to fine-tune your operational framework for the better.

By embracing a culture of continuous improvement and holding a steadfast commitment to ensuring every process reflects your broader ambitions, you set the stage for a future where technology is not simply a tool but a partner in driving success. Taking an integrated approach demonstrates the power of such alignment—it is the bridge between daily operations and visionary leadership, turning fragmented efforts into a harmonious system that supports profound and lasting change.

This deeper exploration of strategic alignment shows that the decision to overhaul existing systems is rooted in a desire to fix gaps but to realize the full potential of your business. It is about creating a mosaic where every piece, every process and every insight contributes to a single, coherent view—one that embodies the aspirations of your organization and drives forward its competitive edge in an ever-evolving marketplace.

Monitor Key Performance Metrics

Throughout this process, it is critical to remain anchored in measurable objectives. The transformation from a fragmented, outdated system to a cohesive, integrated one is best assessed by looking at specific performance metrics. For many organizations, these metrics might include improved cycle times, increased accuracy in forecasting or a reduction in operational costs.

One practical technique that I have found invaluable is setting up a series of short- and long-term benchmarks. These benchmarks are discussed candidly during planning sessions and serve as checkpoints throughout the implementation process.

In one noteworthy example, a manufacturer established clear targets for order fulfillment times and inventory turnover. As the new system rolled out, regular reviews showed progressive improvements, which in turn helped to secure further investment and enthusiastic support from employees across the board.

Leadership

A critical element of this strategic realignment is the role of leadership in continuously guiding and reinforcing the connection between the ERP system and the company's broader ambitions. Leaders must communicate that the transformation is not about chasing a technological trend but about building a resilient operational framework that supports the organization's mission.

This often involves investing in change management initiatives, dedicating resources to post-implementation reviews and even redefining roles to better suit the realities of an integrated system. I have observed situations where, after the primary rollout of a new ERP solution, leadership convened dedicated cross-functional teams to monitor system performance and recommend ongoing improvements. This maintained momentum and created a sense of shared ownership, which is a key ingredient in making the transition a lasting success.

Moreover, embracing iterative learning enhances the alignment of technology with strategic objectives. Rather than expecting a flawless transition overnight, it is important to adopt a mindset of continuous improvement.

Well-designed ERP systems come with powerful analytics and reporting tools that can illuminate areas that may still need fine-tuning. The data they produce can be used to drive further process enhancement.

For instance, in a manufacturing setting I once assisted, the initial implementation led to a surge in production efficiency, but it also uncovered unexpected delays in quality control. By using the real-time analytics available, the company was able to identify new bottlenecks, refine its production schedule and achieve a balance that propelled overall performance to new heights.

This dynamic process of feedback and refinement makes it clear that an ERP system is not a static solution; it is a living component of organizational strategy.

Preparing for ERP implementation is as much about managing change as it is about managing technology. The period leading up to a system rollout is often filled with uncertainties and the inevitable apprehension that accompanies meaningful change. Strategic vision must be integrated with hands-on preparation so that every stakeholder feels confident and ready for the new way of working.

As you prepare your organization for ERP implementation, consider this phase as the vital foundation upon which all future success rests. Every decision made in this stage—from selecting the core team and mapping out new workflows to establishing contingency plans for potential setbacks—has the power to either facilitate a smooth transition or create obstacles that may persist long after the system is live.

For example, I recall a scenario where a well-meaning executive team skipped an in-depth process review, assuming that their longstanding workflows would integrate with the new software. When the system went live, a lack of alignment between the system's workflow and existing processes caused significant operational disruptions and even led to temporary shutdowns in production. It became clear that taking the time to perform meticulous preparations would have prevented these issues, underscoring the necessity of a methodical, all-encompassing approach.

Landing an Ideal Implementation

In many cases, I led projects with meticulous planning, clear communication, and strategic decision-making that had a positive result.

For example, an ERP implementation for a medium-sized manufacturing client went live in a record 93 days. This wasn't just a triumph of technology, but of resolve, coordination and smart problem-solving. Each team member—from the IT professionals to the end users—understood that this was not just a chance to adopt new software but also an opportunity to rethink and reshape the way the business operated.

That success resulted from aligning people, processes and technology to address both immediate concerns and long-term objectives.

Through all these experiences, a pattern emerged: the success of an ERP implementation rests not on fads or fleeting trends, but on a deep understanding of the inherent challenges and a commitment to addressing them head-on.

The contemporary business world is experiencing intense pressure to keep pace with rapid technological and operational changes. In this setting, an ERP system is no longer a tool just for large organizations. It has become a practical necessity for companies of all sizes.

Whether you lead a small business or a multinational conglomerate, the decisions you take when implementing an ERP system have lasting implications for your organization.

I sat at the table with executives grappling with the intricacies of managing change. I have witnessed the transformative possibilities that arise when leaders are not afraid to tackle tough questions head-on. The aim here is to equip you with the technical understanding of ERP systems and the managerial insights that will help you steer the project towards a tangible, meaningful outcome.

Assembling the Right Implementation Team

When planning an ERP implementation, the first critical step is knowing that you have a strong, dedicated team whose expertise and commitment will drive the project forward. The success of this complex undertaking is not solely determined by the software or the budget—it rests on the people behind the scenes who understand both the technical requirements and the operational realities of your organization.

In my many years of guiding organizations through ERP implementations, I have repeatedly witnessed that a well-assembled team forms the backbone of a project that meets its objectives and positions the company for sustained operational improvement.

As you consider your approach, it is essential to view the task of forming an implementation team as the foundation for a long and challenging effort.

Suppose you want to go on a cross-country trip through the backcountry. You will need the right vehicles, a clear map and, most importantly, skilled drivers who understand the route and the terrain. Similarly, an ERP project demands professionals who can manage technical complexities, understand management imperatives and communicate effectively with stakeholders across the organization.

Each team member's role must be clearly defined, with responsibilities covering technology, process oversight, user adoption and ongoing support. When everyone knows their role, the team functions as a cohesive unit capable of overcoming unexpected obstacles.

I recall the situation at a manufacturing firm as it started a major ERP project. The initial plan was optimistic. It was a rapid rollout of an integrated system to consolidate fragmented processes. However, as the project progressed, it became apparent that the team was missing a critical link—a representative who understood the operational nuances of the production floor. With only a technology-centric group in place, the project encountered resistance from employees who felt that their daily challenges were not completely recognized.

The turning point came when a respected production manager was brought on to the team. His insights into the real-time inefficiencies on the shop floor became invaluable. He helped shape a more realistic timeline and built bridges between the technical team and the operational staff. His inclusion turned the project around, helping to ensure that system configurations reflected the everyday challenges and opportunities, rather than just high-level business functions.

Availability of Team Members

An essential aspect of this preparation is forming a dedicated implementation team. The team needs to be free of other obligations to concentrate on the implementation of ERP. The success of any ERP project hinges on having the right people in place. You want those who understand the technical aspects and are deeply acquainted with your organization's operational nuances.

Most of the time, functional leaders of the respective departments are also assigned to perform as ERP implementation team members. However, there is a need to recognize that such functional leaders, who are not willing to extend their work hours during the implementation, can't be effective contributors. This dependency impacts the implementation timeline and budget. Any unavailability of such critical contributors increases the project risks. It helps if their operational responsibilities are realigned to make time for the implementation.

I remember working with a CFO who, despite advocating for a comprehensive system overhaul, was reluctant to invest time in formally assembling a dedicated team. Later, when the team was assembled, none of them had more than 20% of their work time available for the implementation. The project eventually began to falter because conflicting interests and a lack of clear accountability hampered the decision-making process.

In another instance, I led an ERP initiative that brought together a cross-functional team of seasoned professionals from the outset and ensured their priority for the implementation tasks. Their collective expertise quickly identified gaps in the initial plan and drafted a contingency framework that prevented potential delays.

These experiences underscore a fundamental truth: the best technology means little if the people driving its implementation are not fully prepared, available, aligned, focused and enabled to take ownership of the process.

Hire an ERP Systems Coach

If this is your first ERP implementation as a leader, do not go at it alone. Hire an experienced ERP Systems Coach. This decision will significantly increase your chances of success, may save you thousands of dollars and reduce headaches. Setting the right priorities in the correct sequence is fundamental, and the coach will guide you and your team throughout the process. Even a seasoned ERP professional can benefit from having a coach as part of the implementation team.

The ERP vendor will promise you the moon and may give the impression that they have solutions for all potential problems. They may say, "You don't need to spend more by bringing in additional resources. The project manager should be able to carry this out successfully with our support." However, this can be far from the truth.

Reality sets in quickly after you sign the agreement, and responsibility shifts to the vendor's delivery team, with whom you will be interacting only after the ERP implementation project starts. Their focus is to complete the implementation in ways that benefit them the most. You need someone experienced on your side to guide the process so that your organization's interests take priority.

The vendor's sales team members who won the contract will move on to their next pursuit. Their delivery team may politely say, "Sorry, we can't attest to what was mentioned during the sales process, but this is not what the ERP does out of the box." Your ERP Coach will help align your priorities with your operational culture, assess the suitability of the solution and give recommendations throughout the project execution.

Skilled Team with Diverse Expertise

Establishing a well-rounded team starts fundamentally with identifying and choosing individuals whose skills align with the various facets of ERP implementation. This includes technical experts who understand the intricacies of software systems, process experts who can map and re-engineer existing workflows, project managers who can maintain reasonable timelines and budgets and change agents who are adept at facilitating training and user support.

The team's real strength lies in its diversity of expertise. Relying on just a standard IT department is not sufficient. Successful ERP implementations require contributions from across the organization. Representatives from finance, operations, human resources and customer service functions should be part of the team, so that every perspective is considered.

Consider the case of a large service provider that struggled with siloed information and fragmented planning strategies. Initially, the project leadership had assembled a team primarily from its IT department, with occasional input from finance. However, when the project hit several roadblocks, it became clear that the voices of key operational stakeholders were missing.

In response, the leadership expanded the project team to include senior managers from core business units. This move enriched the planning process and built trust among the various departments—an essential ingredient for successful change management.

In this instance, the broader team was better able to diagnose process deficiencies and propose solutions grounded in their departments' everyday realities. The inclusion of people with diverse backgrounds made the overall strategy more comprehensive, and the implementation went more smoothly.

The selection of team members should be based on their expertise and on their track record of working in change-intensive environments. Those who have demonstrated resilience in previous projects and have a proactive attitude towards problem-solving are invaluable.

Technical competence often must be paired with an in-depth understanding of the business's strategic goals. Such individuals can bridge the gap between pure system functionality and the practical realities of daily operations, so that the new ERP system supports actual business practices rather than theoretical models.

I learned this lesson during a particularly challenging rollout at a distribution company where the initial team had technical prowess but little understanding of logistics. Once personnel from the logistics department were integrated into the project team, the differences in process flowed more naturally, and initial missteps in data mapping were quickly rectified. Their presence made certain that the technical design closely mirrored operational reality, a lesson that no blueprint could otherwise have imparted.

Establish Clear Roles and Responsibilities

A critical aspect of forming the right team is establishing clear roles and responsibilities from the outset. Every member should understand not just what they are expected to do but also what they can expect from others. This clarity minimizes confusion and helps prevent overlaps or gaps in responsibilities, particularly during surge periods when issues arise simultaneously in different areas.

I have seen projects where ambiguous responsibilities led to delays. For instance, when two departments assumed the other was handling a critical element of data migration, these assumptions resulted in a lack of accountability and a cascade of problems.

A structured team with defined roles functions as a well-oiled machine, where each gear turns in harmony with the others. This synchronization is fundamental to meeting project deadlines and helps the transition to a new system go smoothly.

It is also essential to recognize that assembling the team is not a one-time event but a continuous process of alignment and realignment as the project unfolds. From the initial planning stages through to the post-implementation phase, the team's composition and focus may need to adapt to changing circumstances.

Early on, the focus might be predominantly on technical integration and data migration. As the project advances, however, the emphasis shifts towards deciding on gaps, user training, change management and supporting ongoing operations. Leaders must be willing to revisit the team structure and make adjustments, adding new specialists when necessary or reallocating responsibilities as new challenges emerge.

A commitment to continuous improvement in team dynamics enables an organization to respond proactively to issues as they occur, rather than being forced into reactive measures that can derail the timeline.

External Communication and Partnerships

Another dimension of preparation involves external communications and partnerships. Modern ERP systems are not isolated tools; they require integration with suppliers, customer platforms and even third-party solutions. Preparing for these external linkages is as essential as managing internal processes.

I've encountered cases where organizations underestimated the complexity of integrating with external systems, only to find themselves scrambling to correct data mismatches and communication lags long after the system was launched.

By proactively engaging with external partners during the planning phase—clarifying expectations, aligning systems and establishing protocols—you can preempt many of these problems. This integration manages your supply chain and enhances your overall competitive position by having every link in the value chain operate in tandem with your core system.

Create a Project Governance Structure

Creating a project governance structure also plays an essential role in team assembly. A steering committee, made up of senior leaders from various departments, creates strategic oversight for high-level decisions that affect the entire project. Their involvement bridges the gap between the implementation team's activities and the organization's broader strategic objectives.

Establishing a dedicated steering committee means that critical issues are escalated promptly and that there is clear accountability for the project's progress. Such a committee meets periodically to review progress against predetermined milestones and to help resolve any interdepartmental conflicts that might arise.

This governance framework, supported by detailed documentation and transparent reporting, reassures stakeholders that the project is being managed effectively and that decisions are made with an understanding of their operational and strategic implications.

Develop a Training and Support Plan

As you progress in organizing your team, another key area to consider is the development of a training and support plan that accompanies the team's formation. The most technical group in the room is ineffective if employees across the organization are not prepared to adopt the new system.

Early in the implementation phase, it is wise to identify champions within various departments who can act as trainers and support staff once the system is live. These champions, often drawn from those who are naturally enthusiastic about working with new technology, can serve as a bridge between the project team and the user community.

Their experience, shared in a language that resonates with their peers, can significantly accelerate user adoption and help smooth out any initial frustrations.

I have seen situations where the lack of such champions extended the learning curve and increased the likelihood of error. In contrast, organizations that proactively invested in a network of informed, accessible trainers reaped the benefits of faster system uptake and higher overall satisfaction.

Encourage Vital Communication

Investing in communication is an equally significant component of building the right team. Regular meetings, clear reporting structures and open feedback channels are indispensable for maintaining alignment.

I have participated in projects where weekly check-ins seemed like routine procedures but turned out to be vital moments when key decisions were made. These meetings provided a forum for tracking progress and for discussing unforeseen challenges, brainstorming solutions and tracking accountability across the board. They allowed each department to voice its concerns and share its progress, creating a shared vision that permeated every level of the organization.

In one case, the consistent dialogue between the finance and operations teams helped identify discrepancies in cost reporting early enough to adjust and prevent a significant budget overrun. Such communication reinforces the notion that the implementation team is not just an administrative body but a strategic partner in driving the transformation.

When forming an implementation team, it is common to encounter resistance, not necessarily from those who wish to contribute but from those who are wary of change. Often, long-standing employees may feel that the new system will disrupt familiar workflows. It is here that the role of the team leader becomes vital, serving as a mediator and a motivational figure who can articulate the strategic vision and foster trust.

In another project I worked on, a senior executive took the time to have individual discussions with key departmental heads to outline the benefits of the ERP system, address their concerns and highlight opportunities for personal and professional growth. This direct engagement was critical in easing the transition and illustrating that the project was not about stripping away existing strengths but about building on them to create a more efficient future.

Continuous Learning and Performance Reviews

The path to assembling the right implementation team is paved with continuous learning and adaptation. As roles and responsibilities evolve, team members may need additional training or support to meet emerging challenges. Regular performance reviews, both at the individual and team levels, can identify areas where further investment in skills is necessary. These reviews are also opportunities to celebrate small wins, a critical practice that reinforces the team's commitment and builds momentum.

When every member sees tangible results stemming from their efforts, from improved process metrics to positive feedback from end-users, it creates a collective sense of purpose that fuels a broader transformation.

In the broader context, forming this team is a strategic investment. The time and resources you devote to building a cohesive group of professionals now will pay dividends later in terms of smoother operations, higher employee morale and the successful realization of your ERP objectives.

The team's work extends beyond installation. It is charged with being certain that the integrated system continuously supports the evolving needs of the business. Their ongoing involvement after implementation, such as post-rollout reviews and ongoing system optimization, makes the ERP solution a valuable asset supporting the strategic needs of the organization.

One of the most essential factors in the success of an ERP project is assembling a well-rounded, multifaceted implementation team. There should be a clear strategy for choosing the right individuals, establishing clear communication channels, defining roles and integrating external expertise where needed.

The stories and examples shared here serve as practical guides to help you identify the potential pitfalls of an under-resourced or misaligned team and to highlight strategies that have proven effective in transforming isolated challenges into structured, coordinated actions.

A Collective Effort

ERP implementation is a collective effort—a partnership among technical experts, process specialists and change agents—all working in unison to reimagine and rebuild how the organization operates. By investing in the right team, you are not just filling positions; you are laying the cornerstone for a new operational framework that aligns with your strategic ambitions and drives measurable improvements.

As you move into the practical phase of implementation, let your commitment to building an effective team serve as the beacon that guides every decision, process review and technical configuration.

With a dedicated, well-prepared team at the helm, your organization will be better equipped to manage the challenges and complexities of ERP implementation. This human element—your team's collective expertise, resilience and adaptability—transforms the process from a technical project into a strategic metamorphosis.

The foundation you build today, with the right people in place, will determine the success of the initial rollout and the enduring impact of the system on your operational effectiveness and business growth.

As you reflect on these insights, consider how the shared experiences can be applied to your own planning process. Evaluate the strengths of your current team, identify potential gaps in expertise and create strategies to bridge those gaps. Understand that the transformation you seek is rooted in your people's ability to work together efficiently and align every decision with the larger goals of the organization.

The work of assembling this team is, in many ways, the first tangible step toward a future where your organization operates as a unified, agile unit—one that turns challenges into opportunities and achieves strategic success through coordinated action.

Now, with the fundamentals of team formation firmly in place, you are ready to move on to the next phase of preparing for ERP implementation. This involves setting realistic timelines and budgets and conducting a thorough review of your business processes to be certain that every operational detail is aligned with the new system's capacities.

As you transition further into the practical aspects of preparation, let the solid foundation you have built with the right implementation team be the engine that drives your organization forward into a future of operational excellence and strategic growth.

Set Realistic Timelines and Budgets

Beyond assembling your team, setting realistic timelines and budgets is equally crucial for the implementation's success. Leaders can be tempted to adopt overly optimistic schedules, particularly when the promise of rapid transformation is on the horizon.

However, my experience has repeatedly shown that shortcuts in planning invariably lead to unexpected overruns and operational disruptions.

Take, for example, the story of a large enterprise whose timeline was compressed in a bid to show results to investors quickly. Soon enough, unanticipated challenges cropped up—ranging from data migration issues to resistance from front-line teams—forcing the organization to extend the schedule and inflate the budget significantly.

By then, the initial enthusiasm had dwindled, and confidence was shaken. The lesson is that a pragmatic approach, which considers all potential variables, is essential to avoid creating a situation where the project is constantly facing emergencies rather than moving forward according to plan.

Setting realistic timelines and budgets stands as one of the cornerstones of successful ERP preparation. This phase is not simply about assigning dates and numbers to tasks; it is about charting a course that reflects the true scope of the transformation while accounting for the inevitable hurdles that arise in a complex project.

I have seen initiatives that rushed the planning process only to face cascading delays and escalating costs later.

Setting a realistic timeline and well-considered budget is not inherently restrictive, but rather protective measures that increase the chance of success while building in the flexibility needed to manage risks.

It all starts with an honest assessment of your organization's processes and teams' readiness. Before setting deadlines, leaders must conduct thorough assessments that account for the technical elements of the project and the human and procedural factors at play.

In one of my earlier projects, I guided a mid-sized organization through an extensive review of its current workflows, data sources and historical performance metrics. This review was eye-opening—the leadership soon realized that their previous projects had consistently underestimated the time needed for employee training and system testing.

Instead of pursuing an aggressive two-month timeline, the revised plan extended the schedule by a month. In the long run, this adjustment prevented a rushed rollout and provided sufficient time for proper training, testing and iterative improvements. What might have appeared as a delay at first saved the project from becoming a rushed and less-than-successful effort.

The Phased Approach

One method I have often recommended is to segment the overall ERP project into distinct phases and treat each phase as a micro-project. This approach allows for a more manageable timeline with detailed milestones that can be achieved incrementally.

In one large-scale project with a multinational client, we divided the work into phases that addressed different operational domains such as finance, supply chain and human resources. Each phase had its own set of objectives, budget allocations and timelines, which were then integrated into the overall project schedule. This provided clarity to all stakeholders and allowed the project team to learn from each phase and adjust the subsequent ones accordingly.

The result was a more flexible plan that could accommodate shifts without derailing the overall timeline. This phased approach reinforces the idea that establishing realistic expectations is a matter of careful upfront planning and of creating a feedback loop that ensures continuous improvement throughout the implementation cycle.

Understand Interdependencies

Building realistic timelines relies on a detailed understanding of all interdependent components. Every phase of the ERP implementation must be considered, from the initial requirements gathering, software configurations, data migration and integration testing to the final transition and stabilization period.

Often, organizations fall into the trap of using overly simplistic project management tools or, worse, assume that digital solutions will automatically streamline the complex interactions among departments.

I remember a case where a well-intentioned project manager tried to compress the planning phase by combining all the preparatory work into one round of meetings. The result was a superficial plan that lacked detail and did not anticipate the ripple effects of technical delays.

Later, when issues arose in integrating older systems with the new platform, there wasn't a built-in contingency plan to fall back on. This experience underscored the importance of breaking down the entire project into granular tasks and mapping dependencies meticulously. Every task, no matter how small it might seem, should have a realistic duration attached to it, along with designated points for review and adjustment.

This yet again reminds us that, if you have not led an ERP implementation project previously, then understanding the interdependencies when estimating the timeline and budget will most likely become a theoretical exercise with a higher potential of significant variations. Seek external help and as recommended earlier, engage an ERP coach.

External Dependencies

Organizations must also consider external dependencies to build a realistic schedule. In many cases, ERP implementations involve integration with third-party vendors, service providers or even government compliance requirements. Each of these external factors can significantly affect the timeline.

For example, a company I worked with experienced delays when waiting for an external service provider to deliver critical software patches. The initial plan had not accounted for this dependency, resulting in a bottleneck that rippled through the project schedule.

By incorporating external dependencies into a comprehensive risk matrix, the team was later able to negotiate better service-level agreements and adjust expectations accordingly. This proactive management of external factors is key to creating a planned timeline that remains realistic and achievable.

Engage External Partners and Consultants

Beyond internal dynamics, an often-overlooked aspect of building the implementation team is the involvement of external partners. Many organizations find that partnerships with experienced ERP consultants or vendors create an external perspective that is both unbiased and grounded in broad industry experience. These external experts can help assess your current systems, recommend the best practices and guide the technical configuration to align with your specific needs.

I once supported a company that, despite having a solid internal team, benefited immensely from an external consultant's experience with similar projects. The consultant's input helped identify potential pitfalls that the internal team had not considered, thereby preemptively addressing areas that could have led to significant delays. Using external expertise to complement internal capabilities encapsulates a balanced approach that can be critical when facing complex implementations.

Budget Planning

Budget planning deserves a similar level of thoroughness. When repository resources are allocated without fully appreciating the full scope of the project—from hardware and software costs to training, change management and post-implementation support—the possibility of overruns increases significantly.

I have seen instances where unexpected costs, such as the need for additional licenses or specialized training modules, led to a sudden 30% increase in total expenditures. These budget variances are not necessarily signs of mismanagement; instead, they are indicators that the planning phase did not completely capture the complexity of the work ahead.

A comprehensive budgeting process considers past projects, industry benchmarks and the specific challenges that could arise from the current architecture.

In one project, for example, the company established a rolling contingency fund, allocating around 15% of the overall budget to unforeseen challenges. In doing so, the team could address issues promptly without jeopardizing the initiative's overall financial framework. This approach instills greater confidence among stakeholders and transforms the budget from a static number into a flexible tool that adapts to the project's realities.

It is equally important to recognize that setting realistic timelines and budgets is an iterative process. No matter how thorough the initial planning, unexpected challenges are certain to occur. The key here is to have regular checkpoints throughout the project where progress is reviewed against the plan, adjustments are made and lessons from earlier phases are applied to the upcoming tasks.

Consider the analogy of tending to a garden. Even with careful planning and regular maintenance, seasonal changes will require you to adjust your approach. Similarly, in an ERP project, early phases such as data migration and initial integrations may reveal issues that necessitate leniency in the schedule or additional funding. Regular review meetings, performance dashboards and project updates help everyone remain aligned with the shifting realities of the project, so that any deviations from the plan are quickly identified and corrected.

Risk Management

In addition to internal and external dependencies, risk management plays a pivotal role in setting realistic timelines and budgets. Every project, regardless of its complexity, carries inherent risks, such as technical uncertainties, staff turnover, regulatory changes and market volatility.

One effective way to mitigate these risks is to conduct a thorough risk assessment early in the planning phase. This can be achieved by gathering input from a wide range of stakeholders, including department heads, IT specialists and external consultants.

In one project, the company developed a comprehensive risk register that detailed every potential issue, along with likelihood ratings and proposed mitigation measures. This risk register served as a contemporaneous guide for the project team and provided a basis for adjusting both timelines and budgets when high-risk areas were identified.

Knowing where potential problems might arise allows leaders to build buffer periods into the schedule and allocate additional funds as a contingency, so that the plan remains resilient in the face of uncertainty.

Create a Risk Register with Contingency Plans

Risk management is another key element when discussing technological challenges. Establishing a foresight mechanism—in the form of a risk register—helps anticipate potential technical issues and prepare contingency plans. This tool should be comprehensive enough to cover everything from system incompatibilities to unforeseen vendor delays.

In one project, the risk register included potential technical obstacles and strategies for mitigating their impact, such as fallback data recovery plans and emergency vendor support agreements. When a significant technical issue occurred, the pre-planned contingencies were activated immediately, minimizing downtime and preserving trust among stakeholders. Effective risk management transforms potential crisis moments into manageable incidents, reducing the overall negative impact on the project timeline and budget.

How to Use Risk Management Effectively

It is important not to underestimate the role of proactive risk management in handling these challenges. Creating a detailed risk register early in the project can help identify potential issues before they become critical.

In one instance, an organization performed an extensive risk assessment that identified even moderately likely events, such as training delays or a minor vendor miscommunication, and allocated contingency resources accordingly. This comprehensive planning ensured that when a risk did materialize, it was addressed swiftly and with minimal disruption.

Regular reviews of this register create a checkpoint where issues are recognized before they become critical, and contingency measures are activated promptly.

In several projects I have witnessed, this proactive approach helped teams adjust timelines, shift budget allocations and refine processes with minimal impact on the overall project trajectory. The act of continuously monitoring and updating risks improves security and builds a culture where challenges are expected, rather than feared.

The proactive mindset embedded in such risk management strategies allows the project to remain flexible, reducing the overall impact of unforeseen challenges and ensuring that progress can continue even when the path forward is not perfectly smooth.

Communicate Timeline and Budget Frequently

Beyond planning for risks, it is essential to communicate timeline and budget expectations clearly across all levels of the organization. Change, particularly of the scale associated with an ERP implementation, can often elicit anxiety, especially if teams feel pressured by unrealistic deadlines or ambiguous resource allocations.

In one case, I observed that when managers were left in the dark about the evolving budget and timeline, it led to pockets of resistance and a general sense of uncertainty. By contrast, projects that embedded transparency into their planning process—through regular updates, open forums and detailed reporting—helped build trust and kept every team member focused on the common goal. This open communication keeps stakeholders informed and drives collective problem-solving when issues arise.

One practical example that resonates with many organizations is the experience of a global manufacturer that was transitioning to a new ERP system. Initially, the project suffered from tight deadlines that did not adequately consider the extensive data reconciliation required for legacy systems. Once these shortfalls became apparent, the leadership team decided to extend the timeline by several months and increase the budget by a modest percentage to accommodate additional testing and training.

Far from being a setback, this adjustment was communicated openly to all stakeholders, who came to appreciate that the revised plan would save time and reduce errors. The outcome was a smoother system rollout along with a more substantial organizational commitment to the new way of doing business, underscoring the value of building in realistic expectations from the start.

Budget discussions, while at times may be seen as dry numbers on a spreadsheet, have profound implications for the success of the ERP project. They serve as a measure of how seriously an organization is willing to invest in its own future. A detailed budget should clearly itemize the costs of hardware, software customization, training, consulting services and even ongoing support after launch.

In one notable project, the leadership team came prepared with a well-documented budget that showed expected costs and projected savings and return on investment over time. This fiscal transparency provided a compelling case for the ERP investment, enabling stakeholders to see beyond the immediate expenditure and understand the long-term benefits.

In many cases, an ERP project's success lies in its ability to deliver measurable improvements, such as reduced operational costs, faster decision-making and enhanced productivity. A realistic budget serves as a tool for controlling expenditures and a benchmark against which success can be measured post-implementation.

Flexibility is Essential

As you manage these aspects of realistic planning, remember that flexibility remains a key principle. While it is essential to lay out detailed timelines and budgets, it is equally important to remain adaptable. The ability to adjust plans in response to new information or shifting circumstances is fundamental to the success of any complex project.

Rather than viewing adjustments as failures, they should be seen as part of an ongoing process of refinement that strengthens the project over time. In a dynamic business environment, rigidity can be as dangerous as poor planning. It is the adaptive organization—one that continuously revises its plans considering emerging data and feedback—that achieves its strategic goals.

Setting realistic timelines and budgets is not about limiting ambition or accepting inefficiency. It is about creating a clear, workable plan that acknowledges the full scope of the challenges ahead. This process involves a detailed understanding of internal processes, an honest assessment of external dependencies, a proactive risk management strategy and an ongoing commitment to transparency and communication.

The team that manages these areas with diligence is far more likely to avoid the pitfalls of rushed planning and unexpected costs, so that the ERP implementation becomes a catalyst for lasting transformation rather than a source of ongoing frustration.

In the advancement toward readiness, the planning phase for realistic timelines and budgets acts as a roadmap and safety net—one that allows you to move forward confidently while accommodating the uncertainties that come with a large-scale change. When every facet of the plan is scrutinized, every potential bottleneck identified and every risk factored into the budget and schedule, the project transforms from a hopeful idea into a structured, achievable initiative.

As you reflect on the experiences and insights shared here—from the pitfalls of compressed schedules to the benefits of continuous risk assessment—consider how these lessons apply to your own organization. Evaluate your current planning processes, engage openly with the diverse voices within your team and commit the necessary resources needed for your timelines and budgets that are realistic and resilient.

The goal is to create a plan that serves as a living document—one that guides the project through its various stages while remaining open to adjustments as new challenges and opportunities emerge.

Now, equipped with a deeper understanding of how to build and manage a realistic timeline and corresponding budget, you are well-prepared to tackle the next critical component of ERP readiness—a thorough review of your business processes. The lessons learned here are a solid foundation for evaluating your current operations and identifying the precise adjustments needed to make your new system work well with your organizational strategy.

Conducting a Business Process Review

A comprehensive review of your current business processes is the critical element in preparing your organization for successful ERP implementation. In many ways, examining every workflow, procedure and reporting mechanism acts as the final audit to determine if each aspect of your organization is ready to integrate into a unified system.

Just as a builder inspects every brick before constructing a lasting structure, you must closely review the existing methods that drive your operations. This exercise forces an honest look at processes that have stagnated over time, sometimes to the point where outdated practices hold your organization back.

Consider a distribution firm I once consulted with that prided itself on its longstanding procedures for handling transactions. Over time, however, these methods had grown increasingly inefficient. Multiple locations managed similar tasks using different spreadsheets and manual reconciliations that, while familiar, were no longer adequate in a fast-paced digital environment.

The leadership team decided it was time to scrutinize every process from end to end. The process review begins with gathering the information, much of which is hidden in the routine tasks that employees perform every day. The review team formed a cross-functional group that spent countless hours mapping out each step, identifying redundant workflows and pinpointing where delays occurred. The review turned out to be a revelation. What was once considered standard practice had silently become a bottleneck, undermining both efficiency and staff morale.

This review highlighted areas in need of re-engineering and provided a rare opportunity to bring together diverse parts of the organization. Each perspective added valuable insights into how processes could be streamlined without sacrificing the quality of the results.

In working with a mid-sized manufacturer, I observed that teams often resist change simply because they have not taken the time to articulate their existing workflow. It was only after the company's team created detailed process maps that the full scope of inefficiencies became apparent.

For example, one production line was hampered by a series of approvals that stretched the time to complete a single order beyond acceptable limits. In discussing these maps with staff, gaps were identified where technology could replace outdated manual tasks. Removing unnecessary steps led to a shortening of the workflow and to a significant boost in employee satisfaction. When people see the potential for a smoother daily routine, the rationale for broader reform becomes clear.

Critically Examine Workflows

Many organizations continue to rely on legacy processes that, while familiar, no longer serve the current operating environment. This often involves critically examining existing workflows—from procurement and production to sales and customer service—and identifying areas where improvements can be made.

In one memorable engagement, I worked with a company that had long prided itself on its time-tested processes. However, when subjected to a comprehensive business process review, it became painfully evident that what had worked in the past was now inefficient and even counterproductive. The review led to making tough decisions, including the re-engineering of several core processes.

While these changes were initially met with resistance, they eventually paved the way for smoother operations and more effective automation once the new ERP system was in place.

This example reinforces the idea that successful implementation is not about preserving the status quo, but about rethinking operations to match the realities of modern business.

Improving on Legacy Systems

A process review is also the appropriate platform for questioning assumptions that have gone unchallenged, often simply because the feeling is, "This is how we've always done it." In every organization I've worked with, there comes a point when established workflows begin to feel more like relics of the past rather than reliable frameworks for the future.

A pharmaceutical company, for instance, clung to a series of legacy procedures for managing research data even as regulatory requirements and competitive pressures demanded quicker access to real-time information. A detailed review revealed that many of these processes were outdated and actively working against the company's competitive edge.

The review prompted a rethinking of the entire data management strategy, which resulted in the integration of automated data capture, analytics and reporting that better reflected current needs. Addressing entrenched habits is daunting, but it is through this kind of rigorous examination that the path to lasting change is found.

Understanding Core Elements

As you reflect on these core elements—the team, timelines, budgets, process reviews, cultural readiness and external integration—it becomes evident that preparing for ERP implementation is not a singular task, but an intricate mixture of interwoven steps. Each step must be given its due diligence and carefully managed to ensure that the broader strategic goals remain in focus.

The investment of time and resources in this preparatory phase is a critical determinant of long-term success. Going through a detailed planning process gives you the tools to anticipate challenges and seize opportunities, so that your ERP implementation goes smoothly and delivers the strategic benefits that your organization seeks.

This phase of preparation is where you set the stage for sustained transformation. It is where insights from earlier strategic assessments are turned into actionable plans, and where the vision for a unified system crystallizes into steps you can follow and measure.

Challenge the Status Quo

Effective progress in a process review requires a willingness to challenge the status quo and to consider even the most minor inefficiencies as significant opportunities for improvement. Every manual touchpoint in your current system represents a potential upgrade, and every repetitive task may be reconfigured into something streamlined and more responsive to change.

What becomes apparent is that a process review is not a critique but a constructive exploration, one that identifies both what works well and what does not within your current operational framework. In doing so, it creates a dual benefit. It preserves the strengths that have helped your organization succeed while paving the way for enhancements that can spur future growth.

One of the more insightful moments in a process review often comes when the team can bridge the gap between the high-level strategic vision and the minutiae of everyday operations.

For example, I once worked with an international logistics provider that, on paper, had a clear growth strategy. Yet, a review of key processes uncovered that the data handling mechanisms used by regional offices varied widely, leading to inefficiencies and miscommunication at the corporate level.

By mapping out these processes and identifying the disparities, the leadership was able to standardize procedures across all offices, which added conformity to the data and supported the strategic goals of scalability and consistency. This standardization was not just a technical fix. It was a cultural shift that helped unite teams under a common operational framework, thereby making the company's grand vision far more attainable and measurable.

Collaborate to Eliminate Inefficiency

This stage of review is inherently collaborative, and it often reveals that the root causes of inefficiencies are complex, interwoven issues rather than isolated incidents.

In one case, a large manufacturing company discovered that its order-processing system was suffering from inconsistent data inputs. This problem was traced back to various departments entering information in different formats. The data inconsistencies forced additional steps to verify and correct errors, generating delays that rippled throughout the organization.

By closely reviewing their business processes, the leadership team was able to standardize data entry protocols. This change simplified reporting and automated several quality checks that had previously required manual intervention. When these processes were fed into the new ERP system, the impact was immediate. The result was improved accuracy, faster turnaround times and a more cohesive approach to handling information. Such outcomes demonstrate the value of diving deeply into existing practices, not to assign blame, but to uncover the hidden opportunities for improvement.

Perform Regular Reviews

It is essential to understand that business process reviews are not a one-time effort. The environment in which your organization operates is continuously changing, and regular reviews are necessary so that your ERP system remains effective. The initial review acts as the baseline for the new system, but subsequent assessments will be needed to gauge the effectiveness of the changes made and calibrate the system further.

Ongoing process audits, performance metrics and regular feedback sessions allow you to capture the iterative benefits of a living, responsive ERP system.

When employees know that their input is valued and that processes will continue to be refined based on real-world performance, they are more likely to embrace the system and contribute to its ongoing success.

It is not uncommon to discover during reviews that legacy processes, once thought indispensable, no longer serve the organization's objectives.

In one notable case, an organization found that its multi-tiered approval process had become a severe bottleneck. While the process had been instituted for good reasons in the past, rapid business growth and technological advancements rendered it inefficient over time. Through careful reevaluation, the organization dramatically simplified the procedure.

The outcome was a marked reduction in processing time and a renewed sense of empowerment among employees who no longer felt stifled by endless paperwork. The revised process improved operational efficiency and aligned more closely with the strategic objective of agility—a quality essential to competing in the marketplace.

Learn from Experience

The stories and examples shared here are not just recollections from past projects. They are practical reminders of the importance of thorough preparation and the value of approaching complex projects like an ERP Implementation with both analytical rigor and a human touch.

As you prepare to transition into the phase of executing your ERP initiative, remember that the groundwork you lay now will determine the trajectory of your transformation. The blueprint you develop for assembling the right team, aligning processes, setting realistic schedules and accommodating external interactions is what will secure the future success of your implementation.

This period is not simply an administrative exercise. It is the transformation's backbone and the structure that supports your organization throughout the challenging yet rewarding transition to a more integrated and efficient operational framework.

With these principles in mind, you are better equipped to bridge the gap between strategic vision and practical execution. Your commitment to detailed, thoughtful preparation is the foundation upon which this next phase will be built, so that when the ERP system is finally launched, every department, role and process aligns with the strategic goals you have so carefully defined.

This preparation transforms uncertainty into a clear, actionable plan and transforms scattered efforts into a coherent path forward. The steps you take now, no matter how challenging, will transform your organization into a more agile, united and strategically aligned enterprise.

Benefits of a Business Process Review

As you consider the role of a thorough business process review within your ERP preparation strategy, it becomes evident that this effort is paramount in bridging the gap between current operations ("AS-IS") and the future you wish to build ("TO-BE"). It is where the abstract goals of integration and alignment take concrete shape, where inefficiencies are transformed into actionable insights and where every stakeholder gets an opportunity to redefine their role within the larger framework.

This collaborative review is as critical as assembling the right team and establishing realistic timelines and budgets because it sets the stage for a managed transition to a unified system that supports your organization's strategic aims.

The business process review acts as both a catalyst and a confirmation of the broader preparation efforts for ERP implementation. It reassures leaders that every step taken—from forming the right team, to building realistic timelines and setting a flexible budget—is grounded in a clear understanding of how the organization currently operates and where improvements are necessary.

The anecdotes and examples drawn from diverse industries all point to one fundamental truth. Without a detailed, honest assessment of current methods, even the best technological solutions can fail to achieve their potential.

Therefore, the information in this chapter helps you be certain that the foundation upon which your future system will be built is strong, resilient and aligned with your immediate operational requirements and long-term strategic objectives.

The lessons learned from past implementations—such as the need for detailed process mapping, the value of bridging communication gaps and the importance of continually refining procedures after the initial rollout—are indispensable for making informed decisions as you move forward. Each insight is a building block that supports the broader narrative of integrating technology, processes and people into a coherent, efficient operating model.

Having laid this robust foundation through team building, realistic planning and a meticulous business process review, you are now prepared to transition to the next overarching phase of your ERP project. The insights gathered here act as a springboard for tackling the challenges of implementation head-on, confident in the knowledge that each element of your preparation is designed to support the technical integration of a new system and the cultural and strategic evolution of your organization.

The strength of an ERP project arises from this preparation. Reviewing and refining your business processes is time well-invested. It is the key to unlocking an operational future that is streamlined, integrated and ready for the challenges of a competitive market. Welcome the lessons learned here and let them guide you as you move to the next stage, where structural planning and execution combine to create an environment where technology and strategy converge.

As you prepare to transition into the next phase of ERP implementation, keep these insights in mind. The detailed and candid examination of your business processes, combined with the successful assembly of your team and the establishment of realistic timelines and budgets, forms the backbone of your ERP initiative. Embrace this foundation as you move forward, confident that the groundwork laid here is a testament to the rigor of your preparation and the solid foundation upon which future operational excellence will be built.

CHAPTER 4

OVERCOMING COMMON CHALLENGES

CHAPTER 4: OVERCOMING COMMON CHALLENGES

While a comprehensive plan, a dedicated team and meticulously refined processes can set the stage for a successful ERP implementation, reality may not cooperate fully. Even the most well-prepared organizations eventually face unforeseen challenges—technological glitches, cultural obstacles and a variety of human factors—that can derail an otherwise promising project if left unaddressed.

This chapter will help you recognize and manage those common challenges. The focus is on three essential areas: 1) addressing technological challenges, 2) steering organizational change, and 3) overcoming resistance by people that so frequently emerges when familiar routines are questioned.

I have seen that even when every preparatory step is carefully taken, hurdles often manifest just when the system starts to come together.

Addressing Technological Challenges

Managing the turbulent waters of an ERP implementation is much like steering through a shifting storm when the map you once trusted begins to change its contours. The challenges come in various forms—technical glitches that seem to crop up unexpectedly, organizational roadblocks that slow down decision-making and the inherent human resistance to change that can undermine even the best-laid plans.

Let's examine shared challenges in depth, using practical insights drawn from real-world experience as strategies for transforming obstacles into opportunities for growth. The objective is to arm you with the knowledge and mindset needed to tackle these issues head-on, ensuring that your project remains on track despite the inevitable surprises that come with any transformative implementation of an ERP system.

Integration with Legacy Systems

One core difficulty is the integration of legacy systems with modern ERP platforms.

In one engagement with a manufacturing firm, the initial excitement for digital transformation quickly met resistance from the very data that had been stored in outdated formats. The company had maintained separate databases for different departments for decades; each developed without a common standard. When the technical team attempted to merge these systems, calls for incompatibility emerged.

The data was stored in various file formats, and the underlying business rules had become inconsistent over time. This discordance meant that even a straightforward task like data migration turned into a labor-intensive exercise, full of manual checks and custom coding segments. The company invested heavily in upgrading its hardware and software, yet the inconsistency of its legacy data significantly delayed the rollout.

The lesson was that a comprehensive review of the legacy systems—coupled with a clear plan for data standardization—was essential before initiating any migration. Without that preparatory work, even the best ERP system would struggle to operate effectively. Addressing such issues early on requires a multi-faceted approach.

One practical strategy is to create a detailed data inventory, catalog every data source and the formats in which it is stored and assess its quality and consistency. This inventory allows technical experts to identify which data sets require cleaning, transformation or complete reformatting.

In another project I managed, the team spent several weeks developing a data mapping protocol that documented the flow of information from the old system to the new ERP. This protocol was a technical blueprint and a living document that evolved as new discrepancies were discovered. The result was a significantly smoother migration process and fewer interruptions during the critical launch period. Such proactive measures demonstrate that investing time in understanding and standardizing data pays dividends throughout the implementation.

Customization Requirements

Another technological challenge is the customization required to tailor ERP systems to the specific needs of the organization. Off-the-shelf ERP solutions are built to be flexible, yet they often require adjustments for the processes of each business.

I recall a service company that needed a high degree of customization in its financial reporting module. Their standard reports did not capture particular metrics that were essential for internal decision-making. The customization process, however, introduced its own set of problems. The software vendor created a patch that initially worked in a test environment but later produced errors once scaled to the full production environment.

The challenge was not simply debugging the software but aligning the customization with the broader architecture of the ERP system. The technical team had to iterate through several versions of the patch, each time coordinating with the vendor and the in-house developers until a stable configuration was reached.

This experience emphasized that customization is not a one-off task. It is a dynamic process that requires continuous feedback, comprehensive testing and the flexibility to revert to standard features if the modifications threaten overall stability.

Network and Hardware Compatibility

Network infrastructure and hardware compatibility can also create unexpected delays.

In one instance, a company rolled out its ERP project across multiple sites. The head office had state-of-the-art network capabilities, but regional offices were operating on outdated hardware with limited bandwidth. These discrepancies meant that while the system performed admirably at central locations, remote sites experienced significant lags, which in turn affected real-time data processing and reporting accuracy.

The solution required a phased upgrade of the regional infrastructure, combined with the adoption of cloud-based solutions for non-critical functions, ensuring minimal disruption while gradually modernizing the entire network.

This case illustrates how technological challenges are rarely confined to the software itself. They often extend to the broader IT ecosystem, requiring a holistic review of all supporting components.

Integrating Third-Party Systems

Alongside data and customization, another recurring technological barrier is the complexity of integrating third-party systems. Many organizations depend on complementary tools, such as customer relationship management (CRM) software, manufacturing engineering system (MES) or human resources management system (HRMS), which must work with the new ERP platform.

In one situation, a consumer goods company encountered issues when its ERP system failed to sync correctly with an existing inventory management solution from a third party. This disconnect led to discrepancies in stock levels and ordering delays.

In response, the company set up an integration task force with experts from both vendors to create an application programming interface (API) that bridged the communication gap. The process involved extensive testing and regular collaboration meetings, which eventually produced a stable connection between the systems.

The key takeaway here is that third-party integrations require as much focus as core ERP functionalities. Overlooked details in these areas can result in operational hurdles that impact the entire business.

Reengineer to Eliminate Root Causes

While it is essential to address technical setbacks head-on, understanding their root causes also creates the opportunity to re-engineer processes for long-term benefits. Each technical challenge presents a chance to reexamine existing workflows and refine them for optimal performance.

For instance, the earlier-mentioned manufacturing firm's struggle with merging legacy data sparked a complete overhaul of its data handling processes. Instead of viewing the challenge solely as a band-aid issue, the firm launched a parallel initiative aimed at standardizing data entry processes across all departments.

This dual approach smoothed the migration process and led to a lasting improvement in data quality and consistency. When technological hurdles are seen as catalysts for broader organizational improvements, they contribute to an environment that is more resilient and adaptable for the future.

Revise Priorities and Resource Allocation

From another perspective, technological challenges often force organizations to revisit their strategic priorities and resource allocation.

In a notable example, one manufacturing company, after several delays in its initial ERP integration phase, undertook a complete review of its resource distribution among various projects. The outcome was a strategic decision to divert additional resources—both financial and human—to address specific technical deficits.

This reallocation proved pivotal in meeting overall project timelines and ensured that the ERP system was strong enough to handle peak operational demands. Although disruptive in the short term, such recalibrations are often necessary adjustments that underscore the fact that technology implementations are dynamic, living projects, subject to ongoing realignment in response to internal and external pressures.

Accessibility and User Experience

Further analysis of technological challenges reveals that many issues center around user interface design and system accessibility. Even technically sound ERP systems may still fall short if end-users find them difficult to use.

One mid-sized enterprise encountered persistent pushback during the pilot phase because the graphical user interface was too complex for the administrative staff. In response, the vendor and the internal IT team collaborated to redesign the interface, streamlining features and creating a more intuitive presentation.

What resulted was not just a cosmetic update but a fundamental rethinking of how the system could better serve its users. Simplifying the user experience is critical because a user-friendly system is more likely to be embraced, resulting in fewer errors, higher productivity and better overall efficiency.

Vendor-Induced Challenges

I have seen companies that faced vendor-induced challenges where the promised customization of their ERP software fell short, requiring scope adjustments or even the reassessment of vendor partnerships.

In one project, a vendor executive was responsible for significant customization that was insufficient for the client's needs. The dissatisfaction that built over time forced the client to re-evaluate their relationship with the vendor, eventually leading to the decision to walk away. While such decisions are never taken lightly, they remind us that the ecosystem in which an ERP system operates is dynamic.

Preparing for external challenges means carefully selecting vendors with proven track records and establishing clear contractual terms, service level agreements and contingency plans in case external inputs do not align with your strategic objectives.

Assess Regularly

Establish mechanisms to review and, if necessary, recalibrate your implementation strategy as the project unfolds. Have regularly scheduled checkpoints where technical performance, budgetary impacts and change management progress are reviewed. At these junctions, issues can be identified early and addressed before they escalate into larger problems.

Furthermore, the iterative nature of ERP implementations means that current solutions lay the groundwork for future improvements. By embracing a flexible approach that includes continuous evaluations and scheduled performance reviews, organizations are better prepared to adapt their ERP systems in response to shifting conditions within the market and the organization itself.

After rolling out its new system, one logistics company used quarterly performance reviews to measure outcomes against set benchmarks. These reviews acted as milestones for system refinement and enabled the company to continually adjust its processes in alignment with evolving operational needs. Such a commitment to iterative improvement optimizes system performance over time and instills a culture of continuous enhancement throughout the organization.

Communicate to Progress

What ties all these challenges together is the need for continuous, open communication. A project that can adapt to change is one where every stakeholder is kept in the loop, from the technical teams on the frontlines to the executives monitoring high-level progress. Regular review meetings, detailed progress reports and a willingness to pivot when necessary are all hallmarks of an implementation that is prepared for disruptions.

There was an instance involving a multinational firm where ongoing review meetings helped identify technical glitches and training shortcomings early and played an essential role in maintaining team morale. The regular sharing of both successes and setbacks ensured that everyone was on the same page and that solutions could be reached through a collective effort. This practice spread a shared vision and reinforced the idea that, even when issues arise, the organization is united in its long-term goal of a successful transformation.

In another example, after encountering repeated delays in a data integration phase, a company I advised instituted weekly, cross-departmental, stand-up meetings. These meetings served as a forum for sharing the latest issues, brainstorming solutions collectively and ensuring that no single team was left to tackle its problems in isolation.

This level of transparency smoothed the resolution process and helped distribute the sense of accountability throughout the organization. It is a clear demonstration that managing challenges is not the sole responsibility of any single department but a collective endeavor that benefits from the insights and experiences of the entire team.

In many ERP projects, technical setbacks emerge as the most visible challenges—those unforeseen issues that can disrupt carefully laid plans and force a team to reassess its strategy. In this section, let's explore a deeper analysis of the technological hurdles that organizations face, examine their causes and discuss ways to efficiently address them to maintain project momentum.

Over the years, I have seen many cases where what appeared to be minor software glitches or data migration issues eventually grew into significant delays. Recognizing these challenges early and having a plan for managing them can make all the difference.

Organizational Change Management

Organizational Change Management (OCM) is a critical component of any successful ERP implementation, ensuring that business leaders can drive adoption, minimize resistance and maximize the return on investment. Implementing an ERP system is not just a technological shift—it fundamentally transforms business processes, workflows and employee roles. Effective OCM starts with clear communication from leadership, articulating the vision and benefits of the ERP system while addressing concerns proactively. Engaging stakeholders early, providing comprehensive training and fostering a culture of adaptability help ease the transition. Change champions within the organization can reinforce key messages and support their peers throughout the process. Additionally, continuous feedback mechanisms and post-implementation support ensure that the organization fully leverages the ERP's capabilities.

Leaders who recognize that ERP implementation is as much about people as it is about technology will position their business for long-term success. Furthermore, establishing metrics to monitor change progress and adjusting strategies accordingly can significantly improve overall adoption outcomes.

Incorporating change management plans into the initial project scope ensures dedicated resources are allocated for training, communication and support activities, which are vital for sustained success. A well-structured change management approach ultimately helps embed new systems into the organizational culture, making future transitions smoother and more effective.

Organizational Context

Beyond the immediate technical and human challenges, the broader organizational context also emerges as an essential consideration. Changes introduced by an ERP system often led to adjustments in established hierarchies and operational roles.

For example, companies may experience disruptions when new reporting structures interfere with traditional departments. One enterprise, after rolling out a central ERP solution, found that departments were initially in disarray as roles and responsibilities became blurred.

Senior managers took proactive steps by establishing clear performance metrics and redefining workflows that aligned with the new system. Frequent meetings helped clarify these new boundaries while providing reassurance that the changes were intended to foster greater collaboration and efficiency.

The willingness to adapt organizational structures, combined with clear communication, enabled the company to restore order and eventually thrive under the new system's framework.

Restructuring that accompanies ERP implementation can disrupt established hierarchies and communication channels. Changes in reporting lines or the introduction of a new centralized database can unsettle departmental boundaries, triggering power struggles or simply confusion about roles.

When one project I oversaw led to a reorganization of several departments, there was an initial period of disarray as teams scrambled to redefine their responsibilities. Senior managers held frequent meetings to clarify these roles, making sure that every employee understood how they fit in.

By outlining clear performance metrics and linking them to the strategic objectives of the ERP system, the management team was able to restore order and gradually realign the organization. This process requires a delicate balance, ensuring that the restructuring is seen not as a loss of identity for any team but as an opportunity to work more collaboratively and efficiently.

Organizational Culture

Alongside technological and human challenges, an organization's broader cultural dynamics play a significant role in the outcome of ERP implementation. Organizational change management isn't simply about introducing a new set of software features. It's about reshaping how every department interacts with information and with each other.

In many projects, I observed that one of the key challenges arises from the cultural inertia that exists within organizations. Employees accustomed to legacy systems may view new processes with skepticism, perceiving them as threats to their established workflows.

Addressing these concerns requires proactive change management efforts—clear communication, comprehensive training programs and a genuine demonstration of the new system's benefits. When employees become active participants in the change process, resistance diminishes and the path to successful implementation becomes smoother.

For instance, I once worked with a company whose shift to a unified ERP platform triggered a major realignment of responsibilities and reporting lines. Initially, this caused confusion and anxiety—people questioned their roles and the entire framework within which they made decisions.

However, with carefully structured communication plans and systematic training programs, the organization eventually adapted. Over time, the new structure led to more transparent workflows and encouraged collaborative problem-solving, which in turn delivered measurable improvements in performance.

Though initially disruptive, this cultural shift proved to be a catalyst for enhanced teamwork and a more agile approach to decision-making.

Challenges Arise Despite Confidence

It often happens that when the technical teams are most confident in their ability to deploy a new system, the first signs of trouble begin to emerge—delays in data migration, unexpected glitches in the software integrations and other technical setbacks that were simply not anticipated during the planning phase.

I remember working with a manufacturing firm that had, by all accounts, prepared diligently for its ERP rollout. They had mapped out every phase of integration and built in contingencies for various scenarios; however, when it was time to merge legacy databases with the new system, a series of compatibility issues surfaced.

The data, stored in formats that had gone unchallenged for decades, refused to reconcile cleanly with the modern coding environment. This resulted in repeated delays and sparked a wave of frustration among the technical staff, who had prided themselves on precision.

Rather than viewing these setbacks as a measure of failure, the firm's leadership chose to see them as learning opportunities. They doubled down on testing, re-evaluated the data mapping processes and brought in external experts who gave fresh perspectives.

The experience underscored a vital lesson, which is, technical setbacks are not an indictment of your overall strategy, but rather an inevitable part of complex systems integration that can be managed with flexibility and a willingness to adjust.

At the same time, technological challenges rarely exist in a vacuum. They are often intertwined with the broader organizational context in which the ERP system is being deployed.

People Management

Technology challenges during an ERP rollout often extend beyond technical fixes to include people management. Even the most sophisticated solutions require buy-in from the technical team, who must troubleshoot, test and often adapt on the fly. The atmosphere in a project under stress can become one of reactive problem-solving, where deadlines loom and frustration mounts.

The role of leadership becomes paramount in such high-pressure environments, so that technical teams are not overburdened, recovery plans are in place and issues are openly communicated with all stakeholders.

In one high-pressure project, after a series of unexpected integration failures, management decided to institute daily briefing sessions. Each meeting served as a forum for discussing progress, identifying problematic areas and brainstorming potential fixes.

This routine fostered a sense of shared purpose among team members and enabled quicker decision-making. When issues are addressed collectively, the burden does not fall on a few individuals, and a culture of collective problem-solving is reinforced.

Cross-Functional Oversight Committee

One method I have seen work effectively is the establishment of a cross-functional oversight committee. Such a committee—composed of technical experts, change management representatives and department heads—serves as a central hub for monitoring the progress of the ERP project.

Their role is to manage internal challenges (like workflow misalignments and employee resistance) and external challenges (such as vendor issues and market disruptions) in a coordinated fashion.

This committee meets regularly to assess whether the strategies in place remain effective in the face of evolving challenges and to decide if adjustments are necessary. With this structure in place, you create a feedback loop that minimizes disruptions and reinforces the commitment of everyone involved in the initiative.

Open Communication and Detailed Documentation

Another powerful approach to managing challenges is open communication and detailed documentation. When a technological challenge occurs, such as a delay in data migration, its cause, impact and planned resolution are documented and shared with all stakeholders.

This transparency builds trust and creates a repository of knowledge that can be valuable for future projects. After all, every organization faces setbacks. What differentiates a successful implementation from one that struggles is the willingness to learn from these challenges and use them to drive continuous improvement.

External Consultants

In addition to fostering internal communication, external partnerships can play a vital role in mitigating technological challenges. It is not uncommon for internal teams to reach a point where external consultants and vendors bring fresh perspectives.

I recall one project where an external consultant was brought in after the internal team exhausted traditional troubleshooting measures. The consultant's fresh perspective, grounded in extensive experience across multiple technologies, uncovered an overlooked vulnerability in the system's integration with legacy databases. That insight allowed for a targeted correction that resolved the immediate issue and led to a broader revision of the implementation plan that would prevent similar problems in the future.

In another project, the in-house team had exhausted many conventional solutions to a persistent integration issue. Bringing in an external consultant with a deep background in both ERP systems and Product Lifecycle Management (PLM) systems created the breakthrough that was needed. The consultant's approach involved re-evaluating the integration architecture rather than patching symptoms, which led to a solution that lasted well beyond the initial implementation period.

These experiences reinforce the idea that leveraging external expertise, when necessary, is not a sign of weakness but a strategic decision to enable the project's long-term success. This is an acknowledgment that complex systems benefit from a diversity of thought and experience.

Manage External Factors

It is also critical to recognize that these challenges extend beyond the confines of a single organization. External factors—vendor issues, supply chain disruptions and shifts in market conditions—can dramatically impact ERP implementation.

One enterprise based its schedule on the timely delivery of software enhancements from its vendor. When the vendor encountered internal issues, the delay rippled across the entire project, forcing the company to re-negotiate timelines and reassess its own internal milestones.

Such experiences reinforce the importance of maintaining a buffer in your project timeline and having contingency plans in place. They also highlight the necessity of choosing vendor partners with proven track records and clear communication practices. When external obstacles arise, they often compound internal challenges, making it imperative that the overall strategy can absorb shocks and adapt as needed.

Synergistic Challenges

While technological issues, human resistance and external factors are significant, there is one other dimension that demands attention. This issue is the synergy between the different challenges. Rarely do these issues exist in isolation. Technological setbacks can exacerbate human resistance if employees perceive that delays are indicative of a poorly managed project.

Conversely, a lack of change management efforts can slow the resolution of technical issues, as frustrated employees might give less accurate feedback, further complicating troubleshooting efforts. Recognizing these interdependencies is essential to crafting a holistic strategy that addresses every aspect of the challenge.

In practice, this means that while you may have separate teams focused on IT issues and change management, there must be a coordinated effort to review how these areas affect each other. The overall goal is not just to solve challenges individually but to be certain that the solutions work together to support the transformative objectives of the ERP implementation.

Continuous Learning and Improvement

It is equally important to emphasize the role of continuous learning and improvement as technological challenges unfold. Technology moves fast, and the best systems today might reveal shortcomings tomorrow. Therefore, successful ERP implementation includes provisions for ongoing system optimization.

Regular system audits, user feedback sessions and performance reviews should be an integral part of post-implementation practices.

For example, after the rollout of an ERP system at a large logistics company, it became routine to schedule quarterly performance reviews. These reviews did not aim to assign blame for previous shortcomings but to identify new opportunities for system enhancements.

Over time, the ERP system evolved through these regular updates, adapting to changes in the business environment and technology trends. This continuous cycle of improvement keeps the system current and ensures that organizational processes are refined iteratively in response to emerging challenges.

Adapt to Challenges

Collective experience across these projects emphasizes the critical importance of anticipating challenges and having the tools to adapt to them. Even when not immediately apparent during the early phases, successes are built upon the ability to overcome hurdles with a combination of flexibility, accountability and continuous communication.

This proactive stance ensures that when a problem arises, it is seen not as an insurmountable obstacle but as an opportunity to strengthen the project's foundation. By incorporating contingency plans into every facet—from technical troubleshooting methods to structured employee feedback sessions—you are confident that your ERP implementation remains on a forward trajectory, even when the unexpected occurs.

Progress is measured not by the absence of obstacles but by how adeptly an organization responds to them. Your ability to anticipate and manage technical setbacks, cultural shifts and resistance from within will determine the success of your ERP implementation. Draw on the experiences of others, lean on the strength of your team and maintain the focus on your strategic objectives.

Setbacks are Growth Opportunities

A further consideration is the importance of fostering a culture where challenges are seen as inevitable steps in the process rather than insurmountable roadblocks. Leadership plays a pivotal role in setting this tone.

By framing setbacks as opportunities for learning and growth, managers can help to dissipate feelings of fear and uncertainty.

In one transformation project, rather than downplaying a series of performance hiccups in the early rollout, the executive team opted to openly discuss what had gone wrong in their quarterly meetings. Instead of castigating the errors, they highlighted the corrective measures that had been taken and set new benchmarks for future performance. This approach diffused potential guilt or blame and served to boost morale by showing that the organization was committed to learning and continuous improvement.

When employees observe that every challenge is met with constructive analysis and immediate action, they become more engaged and invested in the project's overall success.

In reflecting upon these challenges, one can see that each setback reveals an aspect of the broader system that can be improved.

For instance, when integration issues delayed a data consolidation process, it prompted a deeper look into legacy systems and an eventual upgrade that solved the immediate problem and enhanced overall system performance.

Similarly, when staff resistance was encountered, the subsequent training sessions and revised communication plans accelerated the adoption of the new system and strengthened the bonds of trust between the workforce and leadership.

In another example, when technical difficulties slowed data integration, a company did not simply try to apply a temporary fix; instead, they revisited and refined their entire data processing strategy, standardizing procedures that strengthened their overall system performance. When employee resistance surfaced, a deliberate shift in communication and training strategies resolved the immediate concerns and built long-term organizational resilience.

In both cases, the challenges served as catalysts for a deeper understanding of the system's intricacies and drove meaningful improvements that endured well beyond the initial implementation phase.

These powerful strategies—ranging from detailed risk assessments and cross-functional communication to proactive change management and contingency planning—are essential tools that can help transform challenges into opportunities for further refinement and success. Managing these familiar challenges is not about avoiding adversity altogether but about building the adaptive capacity to thrive in its presence. With the proper mindset, strategic planning and the collective effort of a dedicated team, each hurdle you encounter serves to enhance the strength of your new ERP system and solidify your organization's operational integrity.

The cumulative experience of managing these challenges will, in time, build an organizational resilience that extends well beyond the ERP project itself, positioning your business for lasting success in an ever-evolving marketplace.

Reflecting on the examples and lessons shared throughout this phase of ERP implementation, I recall a manufacturing firm that, despite comprehensive planning, encountered severe data integration issues, which required an agile response.

There was also the company that managed to overcome internal resistance through decisive, empathetic leadership and effective training—it transformed initial skepticism into active advocacy for the new system. Then, there were projects where external factors like vendor performance and market shifts necessitated rapid adjustments to both budget and timeline.

Each of these scenarios highlights that the path to ERP success is rarely straightforward. It is a challenging process that tests the resilience of both the technology and the people behind it.

Success in Organizational Change Management (OCM) for ERP implementation can be measured using a combination of qualitative and quantitative indicators. Key metrics include: 1) user adoption rates, which track how effectively employees are utilizing the new system, 2) engagement and feedback scores, gathered through surveys or group meetings to assess sentiment and readiness and 3) training completion and competency levels, ensuring employees have the necessary knowledge and skills.

Additionally, measuring business process efficiency before and after implementation provides insight into improvements, while tracking resistance levels—such as the frequency of help desk tickets or the number of workflow disruptions—can highlight ongoing challenges.

Another crucial measure is return on investment (ROI), which looks at whether the ERP system delivers anticipated business benefits, such as cost savings and operational improvements. A well-structured OCM approach ensures leaders proactively address concerns, support employees through the transition and create a resilient organization that fully leverages their ERP capabilities.

Overcoming Resistance to Change

It is essential to reflect on the multifaceted nature of the challenges that inevitably arise during ERP implementation and the strategies that can turn these obstacles into success points. Technical setbacks, human resistance and external influences are interconnected factors that, when managed proactively, contribute to a successful rollout and to your organization's long-term resilience.

Consider, for instance, the process of integrating legacy systems with modern ERP software—a process that has repeatedly shown us that the details matter.

A manufacturing firm once struggled with archaic databases that stored data in numerous inconsistent formats. The technical team attempted to merge the old and new systems, which initially caused significant delays. Rather than allow these issues to derail the project, the team initiated a comprehensive data inventory and mapping exercise that uncovered the immediate incompatibilities and systemic patterns of outdated practices that needed a modern approach.

The solution, which included creating standardized data protocols and engaging external experts, transformed what started as an overwhelming technical challenge into a structured, manageable process.

This example reminds us that detailed preparation, careful analysis and a readiness to adjust strategy are indispensable in overcoming technological hurdles.

Human Behavior

Technical challenges represent only one aspect of this complex landscape. Human behavior, too, plays a critical role. Every employee, from the front-line operators to middle management, interacts with the ERP system daily. Their views, aspirations and most importantly, their opinions must be acknowledged throughout to make a successful implementation and also to chart the least resistive path forward.

In one notable case, a long-standing manufacturing company was hesitant to shift away from familiar methods, with many staff members voicing concerns over a system that could potentially render their expertise obsolete.

Recognizing these fears, leadership took the time to hold one-on-one sessions and focused on targeted workshop training. They illustrated how the new system would improve daily activities and support existing roles rather than replace them.

Over time, this cautious approach helped build internal confidence and gradually shifted opinions. The success of that project was not measured solely by technical milestones but by a cultural metamorphosis where skepticism was transformed into collective ownership.

This shift underscores a key lesson, which is, when people are part of the process and their concerns are addressed transparently, resistance diminishes and overall implementation gains a firmer footing.

Why People Resist Change

I vividly recall a situation where a company, despite investing heavily in technology, struggled with low employee-adoption rates. Front-line workers, long accustomed to specific methods, viewed the new system with suspicion.

In one particularly memorable case, I witnessed a seasoned team member express concern that the ERP system would render his hard-earned expertise obsolete. Instead of treating this resistance as a setback, leadership took it as an opportunity to initiate one-on-one sessions, explain the long-term personal benefits and improved work conditions that the system would bring and gradually help him see that his expertise was still essential—to guide the new system rather than be replaced by it.

This scenario highlights how human resistance often stems from fear of the unknown rather than a substantive objection to change. Managing these challenges requires sensitivity and a proactive change-management strategy that actively involves employees at every level.

Training sessions, internal forums and clear communication about how the new system will enhance rather than diminish employees' roles are critical. Change, when induced from within, facilitates proactive participation and delivers long-lasting results. When employees feel heard, engaged and understand that the ERP system is a tool designed to support their everyday efforts and improve efficiency, resistance can diminish significantly. This increases overall buy-in, ensuring that the system is not just launched but embraced as part of the organizational culture.

Dealing with Fear of the Unknown

Dealing with "fear of the unknown" can be challenging, as it often stems from uncertainty and a lack of control over future situations. It is a common human experience that can evoke anxiety and hesitation, making it difficult to move forward. However, by developing coping strategies such as gathering information, proactively communicating the "what, why and how" and demonstrating the benefits, leaders can learn to manage and reduce this fear. Embracing the unknown as an opportunity for growth rather than a threat can also help transform your perspective and build resilience in uncertain times.

I recall a situation with a company in the manufacturing sector where seasoned staff were initially resistant to the new ERP system, not because they doubted its potential benefits, but because the prospect of learning unfamiliar processes created a fear of the unknown. Employees found comfort in routines that had been used for many years, even if those routines were riddled with inefficiencies.

To address this, leadership initiated a series of informal workshops and one-on-one training sessions. They invited employees to share their concerns and worked to tailor training programs that explained the technical workings of the system and demonstrated how the new methods would make their work easier and more effective.

Over time, the initial skepticism gradually transformed into a measured optimism. This example illustrates that overcoming resistance is more about bridging familiarity with improvement than it is about forcing radical change upon a reluctant workforce.

When Your Best Efforts Go Wrong

In one instance, a well-respected mid-sized enterprise had methodically planned its ERP transformation. It had an experienced team, realistic budgets and detailed process reviews. Yet, once the technical integration began, it encountered a series of unexpected technological issues that threatened to significantly delay the project.

Data migration proved more complex than anticipated, and integration with legacy systems exposed hidden coding conflicts that no one had foreseen. Overcoming these challenges required technical skill and a willingness to rethink assumptions and adjust timelines.

This experience taught me an important lesson, which is, technological challenges are not a sign of failure, but rather an inevitable part of any large-scale transformation. The key is to build both flexibility and resilience into your strategy from the outset so that when these issues arise, you can address them without losing sight of your overall goals.

In practice, this means establishing strong testing protocols, setting aside contingency funds and adopting a decision-making process that can respond rapidly to technical setbacks. It also means fostering open communication between IT teams and business leaders, so that technical challenges are understood in the context of operational priorities. Transparency in sharing the efforts to overcome these hurdles ensures that corrective actions are taken quickly and that all stakeholders remain aligned on the project's strategic objectives.

Beyond the technical domain, the human element frequently emerges as one of the most challenging aspects of ERP implementation. Change, especially the kind that disrupts long-established workflows, is invariably met with skepticism.

Align People and Technology

Equally compelling is the challenge of aligning people with the technology. The ERP system is only as effective as the people who use it, and often, human behavior is the most unpredictable variable in any implementation. Resistance can manifest as reluctance to adopt new procedures, a temporary decline in productivity as employees adjust or even active pushback against the changes proposed by management.

In one instance, a logistics company encountered significant delays because its warehouse staff were slow to move away from the conventional methods they had trusted for decades. Recognizing that skill gaps were at the heart of the issue, management launched an intensive training program that paired tech-savvy trainers with experienced staff. They also established a feedback mechanism, allowing employees to report issues and suggest improvements in real time.

The alignment between seasoned employees and modern technology fostered a more collaborative environment, where the system was no longer seen as an alien force but as an evolving tool designed to support them. This strategy of combining training with active feedback channels is essential for mitigating fears and ensuring that the ERP system evolves in alignment with the users' needs.

External Influences

It is also essential to recognize that many of these challenges are not isolated incidents but part of a larger ecosystem that includes external influences. Vendors, supply chain partners and market conditions all play roles that can affect the progress of an ERP project.

For instance, a manufacturing company once faced delays because its technology vendor had vested interests in recommending one of their colleagues for the role of CIO, when that position opened up during the initial stage of the implementation. The ripple effects from that delay necessitated a re-evaluation of vendor relationships, leading the customer to renegotiate service terms and secure a better, more reliable partnership.

This experience taught the organization that external influencers must be managed appropriately. Establishing service level agreements, maintaining clear lines of communication and having contingency plans in place are not just formalities; they are critical components of a resilient implementation strategy.

Proactive Management of Challenges

Through these varied examples, your attention is drawn to a central theme, which is, challenges during ERP implementation are multifaceted, and success depends on a proactive approach to managing them.

Rather than treating technological setbacks as isolated incidents, successful organizations recognize them as part of a larger pattern that requires both technical and organizational adjustments. Similarly, addressing human resistance is more than just providing training—it involves transforming the organizational culture to make change an accepted and even welcomed part of the business environment.

Managing these shared challenges is not an endpoint but a continuous process. As you progress through your ERP implementation, you will likely face novel issues that were not anticipated during the planning phase. What remains critical is that your organization has the resilience and flexibility needed to adapt.

As you reflect on these detailed strategies and experiences, it becomes clear that the challenges encountered during an ERP implementation are multifaceted yet surmountable with the right approach. The collective lessons of managing technological setbacks, addressing human resistance, managing external dependencies and fostering open communication all converge toward a common outcome to create a more resilient, effective and integrated operation.

Drawing these insights together, here are the key takeaways on managing these challenges:

- Recognize that technical glitches, often seen as threats, are opportunities to refine your data migration and integration strategies, emphasizing the need for standardized protocols and thorough testing.

- Understand that human resistance is natural. Address this through targeted training, open communication and by including key representatives from all affected departments in the change management process.

- Prepare for external factors by establishing clear vendor agreements, building risk registers and incorporating contingency plans that safeguard your project timeline and budget.

- Foster a culture of continuous learning and improvement, where regular review sessions, transparent reporting and external consultations manage every setback with responses that align with long-term goals.

- Emphasize flexibility and adaptability throughout your project, ensuring that each challenge encountered becomes a learning moment rather than an insurmountable obstacle.

These lessons do not exist in isolation. They are integral in creating an ERP implementation that is technically sound, culturally aligned and strategically resilient. The real success of an ERP project is measured not just by whether the system goes live on schedule, but by how effectively an organization is able to absorb the shocks, learn from challenges and continually optimize its operations in response to both internal and external dynamics.

As you move forward with your ERP implementation, remember that every technical glitch, every instance of hesitation and every external disruption is part of a larger process of transformation. The challenges discussed here are universal to complex projects and are as much about people and growth as they are about technical integration. Embracing these challenges, learning from them and using them as catalysts for systematic improvement will help you know that your ERP implementation is successful in the short term and sustainable in the long run.

CHAPTER 5

EXECUTION: THE HEART OF IMPLEMENTATION

CHAPTER 5: EXECUTION: THE HEART OF IMPLEMENTATION

The next step is turning plans into tangible results. With a clear strategy, a well-prepared team and a resilient plan to overcome common challenges, the key is to take decisive actions that transform your organization's vision into a working reality.

This stage is about execution—identifying the right technology partner, aligning the system's capabilities with your established processes and ensuring that data and functionality are integrated seamlessly. This section gives you the practical insights and real-world guidance needed to move from planning and problem-solving into the hands-on phase of ERP implementation.

I recall a striking case involving the CFO of a medium-sized company who had been the project sponsor from the very beginning. After many months of dedicated effort, the ERP project had spiraled into severe delays and budget overruns. Faced with a steep increase in expenses and plummeting morale, the CFO made an urgent call for help.

In a series of frank discussions with top management, we unpacked the expectations and challenges that had built up over time. We then charted a new plan that reined in costs, tightened timelines and realigned the project with the company's strategic goals.

This experience reinforced a crucial lesson, which is, while planning and preparation are essential, it is the execution phase—where decisions are implemented on the ground—that secures the project's success.

Execution is a Dynamic Process

It is also worth noting that execution is not just a one-time event, but a dynamic process that may require continuous adjustments. As the system goes live, there will be opportunities to fine-tune configurations, address emerging challenges and optimize performance through iterative feedback cycles. Each adjustment, no matter how small, contributes to the overall stability and reliability of the platform.

Over time, as the organization becomes more familiar with the new system, these iterative improvements help in locking down the efficacy and strategic alignment that were envisioned right from the start.

A practical approach that has served many organizations well is to break the execution process into distinct phases with well-defined milestones. By segmenting the implementation into clear, achievable blocks, leaders can monitor progress and make informed decisions at each step.

For instance, a phased rollout might begin with a pilot module—such as finance or inventory management—before extending the system to other departments. This allows the organization to test and validate key functionalities in a controlled environment, gather critical user feedback and address potential issues before a full-scale launch.

I have seen multiple instances where a phased approach mitigated risk and built enthusiasm and confidence across the organization as early wins demonstrated the benefits of the new system.

Manage Dependencies

In addition to internal milestones, it is crucial to continuously assess and manage external dependencies during execution. Changing market conditions, vendor performance and supply chain disruptions can all have an impact on your implementation timeline and overall success.

One company I worked with had its rollout affected by delays in vendor deliveries. In response, the leadership negotiated revised service level agreements and adjusted internal timelines accordingly. Being prepared to manage these external factors is part of the strategic agility that defines successful execution. It reinforces the notion that a disciplined approach, supported by contingency resources and flexible planning, is essential to meet immediate goals and to build long-term operational resilience.

Ultimately, successful execution lies in the ability to blend meticulous planning with adaptability. Every technical decision, process reconfiguration and training session contributes to a system that aligns with the strategic objectives set forth in earlier chapters.

The execution phase is where intent meets action. It is the moment when carefully constructed plans are put to the test, and your pre-planned strategy is evaluated against real-world performance. The success you achieve during this phase will set the tone for ongoing system optimization and long-term operational success.

By selecting the right technology partners, balancing customization with standardization, ensuring flawless data migration and maintaining clear communication, internally and externally, you are well on your way to transforming your ERP vision into a vibrant, functioning system.

The steps you take now will mold your ERP system into a tool that meets today's needs and adapts gracefully to future challenges.

Let's explore how an ERP system becomes a tool for continuous improvement once it is fully entrenched in the organization's workflow, serving as a platform that reinforces collaboration, accountability and a culture of innovation.

Communication is the Key

Transitioning from planning to the execution phase calls for a steadfast focus on communication. Every detail, from schedule adjustments to lessons learned from earlier phases, must be shared openly with the team. Regular, structured updates help everyone remain aligned, and potential impacts of newly discovered issues are understood immediately.

I recall the instance of a multinational firm that held daily cross-departmental meetings during a critical phase of data migration. These meetings functioned as a safety valve, allowing team members to express concerns, celebrate small breakthroughs and recalibrate procedures as needed. This level of transparency helped to resolve issues quickly, boosted collective morale and reinforced the strategic vision behind the transformation.

At its core, the execution phase is about turning strategy into practice. All the insights gathered from planning, risk management and process reviews now come together in a coordinated effort to implement the ERP system effectively. It is here that the strategic visions and aspirations outlined in earlier discussions are put to the test, and the true worth of your preparation is revealed.

The decisions made during execution—whether it is selecting the right vendor, prioritizing standard features over heavy customization or ensuring a seamless data migration—will have a lasting impact on your organization's operations.

Expect the Unexpected

Execution also involves addressing the inevitable roadblocks that emerge during the transition. Even the best-laid plans encounter unexpected issues—be it alignment problems between system modules or the staff struggling to adopt new ways of working.

In a manufacturing company I once advised, initial rollouts revealed that employees were having difficulty using the system interface, even though the technology was sound. Instead of forcing the pace, management took a step back to conduct additional tutorials and user experience workshops. This smoothed the transition and created a sense of ownership among staff.

When front-line users see how a tool can simplify their work and reduce manual errors, resistance fades and acceptance increases. These stories illustrate that execution is not only technical; it is fundamentally about people. A project's success is ultimately measured by how those at every level integrate with the new system in their daily routines.

Selecting the Right ERP and Right Vendor

Selecting the right ERP and right vendor is perhaps one of the most consequential decisions you will make during the execution phase. With the groundwork laid through meticulous planning and process review, it is now time to determine which solution will ultimately become the engine driving your organization's transformation. This means establishing the criteria for selecting an ERP vendor, the process for ensuring a good fit and the real-world experiences that illustrate why a careful evaluation process is indispensable.

It is not about choosing the most popular or heavily marketed product. It's about finding a technology partner whose vision, capabilities and approach align with your operational realities and strategic ambitions.

My experience has shown that many organizations fall into the trap of relying solely on executive relationships or attractive sales pitches when selecting an ERP system. I once encountered a case in which a well-connected vendor executive managed to secure an early agreement with a mid-sized company solely based on promises of a groundbreaking solution. However, when the system was finally implemented, it became clear that the functionality covered only about 40% of the client's requirements.

This mismatch was a costly lesson in the dangers of prioritizing relationships over an objective evaluation of the solution's capabilities. The takeaway is clear, which is, a thorough vetting process—which includes independent assessments and hands-on demonstrations—is essential to know that the vendor's offering truly aligns with your business needs.

Process Mapping

Process mapping has a critical role in ensuring successful ERP implementation. This helps businesses document their requirements prior to selecting an ERP system. Quite often, the priority of this step is overlooked and only considered after an ERP System has been selected. Such omissions have negative effects and cause implementation delays and failures.

Process mapping is aimed at delivering the following outcomes as described below, which builds the requirements document to be used to assess a potential ERP Solution.

A. Assessment of Current State

- Document existing workflows, systems and pain points.
- Identify inefficiencies, redundancies and manual processes.
- Gather input from stakeholders to understand operational challenges.

B. Definition of Future State

- Establish clear business objectives and ERP system expectations.
- Determine how the ERP should improve efficiency, automation and reporting.
- Align future processes with industry best practices and compliance requirements.

The following standards and practices have helped ERP implementation projects significantly through the process mapping.

1. Business Process Model and Notation (BPMN)

- A graphical representation for documenting business processes.

- Helps organizations visualize workflows, making ERP integration smoother.
- Supports automation by mapping processes that ERP systems can execute.

2. ISO 9001 – Quality Management System

- Ensures consistent quality in business processes.
- Helps organizations standardize workflows before ERP implementation.
- Reduces errors and inefficiencies in operations.

3. APQC Process Classification Framework

- Provides benchmarking tools for process improvement.
- Helps businesses compare their workflows with industry best practices.
- Supports data-driven decision-making in ERP configuration.

4. ITIL Framework – IT Service Management

- Guides best practices for managing IT services.
- Ensures smooth ERP deployment with minimal disruptions.
- Helps organizations align IT operations with business goals.

5. GAAP and IFRS – Financial Reporting Standards

- Ensures compliance with accounting regulations.
- Helps businesses structure financial data within ERP systems.
- Supports accurate reporting and decision-making.

6. Data Governance Standards (GDPR, CCPA)

- Protects data privacy and ensures compliance with regulations.
- Helps businesses secure sensitive information within ERP systems.
- Reduces risks of data breaches and legal penalties.

7. The SCOR Digital Standard (by ASCM)

- The Supply Chain Operations Reference (SCOR) model gives organizations the ability to assess and improve their supply chain.
- Makes it possible to pinpoint core process areas that need optimization.

Requirements Document

With the definition of the future state of business operations, building a detailed requirements list is the next step forward. This document should also cover technical specifications—such as data processing speed, integration capabilities and system scalability—and address softer aspects like user-friendliness, training support and vendor responsiveness.

I recall guiding a manufacturing firm through this very exercise. Their initial list of requirements was broad and, frankly, overly optimistic. Through a series of workshops involving department heads from production, finance and customer service, they were able to refine their requirements, separating critical functionalities from those that were simply "nice-to-have."

This refined document served as a blueprint during vendor demonstrations to determine that the selected system would be a suitable platform capable of evolving with the organization.

When evaluating vendors, it is equally important to consider the vendor's track record and reputation within your industry. Ask for case studies, references from similar companies and performance metrics that demonstrate past successes.

One of my clients, a fast-growing manufacturing company, initiated discussions with several vendors, but one standout was a vendor who had delivered successful implementations for companies with similar scale and complexity. The vendor's willingness to provide detailed reference cases and even arrange site visits where the ERP system was in operation helped solidify the decision.

These real-world examples give reassurance about the vendor's capability and offer practical insights into potential pitfalls and best practices that might not be evident in a sales presentation.

Choosing the Right Solution

Selecting the right solution is paramount in this execution phase. Even the best vision can falter if the chosen ERP system does not match the unique needs of your business.

Over the years, I encountered situations where project sponsors rush into vendor agreements purely based on executive relationships, only to find that the system's functionality falls far short of their operational requirements. This underscores why a thorough evaluation of an ERP vendor's platform is essential.

Leaders must balance technical specifications with the vendor's ability to match not just one but all facets of your business processes. This means asking probing questions, testing system capabilities in a controlled environment and understanding what the software does today and how it can adapt as your company grows in the future.

Most of the time, this phase of the ERP Implementation initiative is the most arduous and lengthy. The company issues an RFQ to start the selection process. The caution is, if the entrusted leader has not gone through this process successfully before, then there is a very high probability of getting it wrong when this is the first time. In such a case, engage an external consultant or an ERP system coach to assist you through the process, and the cost savings will likely improve the ROI manifold.

Most modern ERP systems meet about 80% of the business requirements of most companies. The known rule of thumb in this context for a mature company has been that about 60% of its requirements can be met by the ERP out-of-the-box, the next 20% of the requirements can be met by configuring the ERP to the closest using preferences/options and the remaining 20% of the requirements will necessitate some short of customization involving software development.

The real challenge starts with selecting an ERP System that delivers the specialized requirements, which may minimize the remaining 20% of your company's customization requirements. Unless guided by an experienced consultant, the selection team may not be able to qualify either the fitment or the feasibility of the ERP vendor's proposed enhancements.

The following are the key characteristics of an ERP system that can help decide whether an ERP software is the right ERP for your organization.

Customization

A fascinating aspect of ERP systems is how they can be tailored to meet the needs of each business. There is no one-size-fits-all solution in the complex ecosystem of modern commerce. Whether a company is a nimble startup or a sprawling multinational, each organization must adapt the core system to its distinctive processes and objectives.

I recall a scenario involving a client in the manufacturing sector whose operations were so unusual that even the best off-the-shelf ERP solution needed significant customization. After a thorough assessment, we determined that aligning the system's configuration with their operational steps was the only way to truly capture the business value.

This process, while time-consuming, proved to be immensely beneficial in the long run. The tailored approach improved operational efficiency and boosted employee morale, as workers found that the system reflected their workflow rather than dictating it. By taking the time to customize rather than conform, the company reaped rewards that went well beyond simple automation.

Scalability

Another essential component of ERP systems is their scalability. In a fast-paced environment, businesses must be nimble enough to adjust quickly to market demands. A rigid system that works well for a modest-sized company might not be suitable as the organization expands.

My experience has shown that leaders who think ahead and design their ERP deployments with scalability in mind are better positioned to manage growth without facing operational bottlenecks.

One such example involved a company that, after a successful ERP rollout, experienced rapid growth due to unexpected market opportunities. Because the system had been designed with flexibility as a core principle, it accommodated the increased volume of transactions and the expanded requirement for real-time data processing. This was a clear demonstration that a future-proof ERP system is not just an investment in current efficiency but also a safeguard against the unpredictable nature of business expansion.

Resilience

Alongside scalability, resilience is a trait that distinguishes successful ERP systems from less effective ones. Unexpected issues can, and often do, arise during transitions, whether due to unforeseen gaps in data, integration challenges or gaps in training.

Resilient systems are engineered to isolate such challenges, allowing the broader operation to continue functioning. I observed situations where; despite encountering significant technical problems during the initial rollout, organizations were able to maintain business continuity because their ERP systems were structured to handle localized failures without cascading into system-wide chaos.

This resilience is not incidental. It is the product of deliberate planning, rigorous testing and a commitment to building a system that can handle challenges over time.

User Interface

Another rich area of discussion is the interface between the ERP system and the end-users. The technology itself must be user-friendly and intuitive. Even the most powerful system will falter if it is not embraced by those who operate it daily.

I recall attending training sessions where skeptics turned into advocates simply because the system was responsive to their needs and easy to use. These sessions were not just technical orientations but conversations that demystified the technology and highlighted its benefits in tangible terms. With a pragmatic "AI-by-the-Side" architecture, the ERP System may offer a choice of user interactions, such as "click" and/or "converse."

The success of an ERP implementation is measured not in lines of code but in users' willingness and ability to integrate the system into their daily workflows. Time and again, I have seen that when the end-user is supported, the entire organization benefits. This is a testament to the idea that technology should be a tool to enhance human capability rather than replace it.

Transparency

Technical complexities notwithstanding, perhaps the most compelling benefit of a well-integrated ERP system is its ability to foster transparency and produce the related and relevant information with a conclusive level of detail. When every data point is connected in a unified system, accountability becomes not just a managerial pursuit but a shared organizational value. The capability of ERP software to let users drill down to the core of the information from their own perspectives is a key differentiator during the ERP software selection process.

Consider the impact of clear, accessible data on operational efficiency. In one instance, a manufacturing plant was grappling with issues of quality control. The delays in identifying defective products were costing the company both in terms of wasted materials and damaged reputation.

With an integrated system that meticulously tracked production metrics, issues were quickly identified, examined and rectified. Process transparency meant that every stakeholder, from the shop floor to the executive suite, was aware of the challenges and could contribute to the solution. This collective ownership of data often leads to improved collaboration and better business outcomes.

Flexibility and Ease of Integration

Another key element in selecting an ERP vendor is the flexibility and ease of integration with existing systems. Even after years of process review and data migration planning, organizations rarely start with a perfectly isolated system. There will always be legacy platforms, third-party applications and external data feeds that need to interconnect with the new ERP solution.

Look for vendors who offer robust integration tools and have a proven methodology for smooth transitions.

For example, I once worked with a client whose chosen vendor had an impressive integration module designed to bridge disparate systems. This allowed the organization to maintain critical legacy functions while gradually transitioning to the new platform, minimizing disruption during the changeover. Such a dual approach—ensuring innovation and continuity—is a hallmark of a vendor that understands the real-world challenges of ERP implementation.

Choosing the Right Vendor

The process of selecting the right ERP vendor requires a disciplined, comprehensive approach that integrates technical evaluation, cost analysis, support assessment and cultural alignment. The decision is too significant to be rushed and must be approached with a long-term perspective that goes beyond initial implementation.

The anecdotes and experiences shared here underscore that the right vendor will deliver a powerful, flexible and integrated system but will also stand by your organization as a committed partner throughout the transformational journey.

As you embark on this critical selection process, keep these four key principles in mind: 1) develop a clear, detailed requirements document, 2) assess the total cost of ownership rather than just upfront expenses, 3) evaluate integration capabilities and customization potential carefully and 4) be certain that the vendor has a strong support system and a forward-looking style.

Engage in thorough due diligence by gathering references, seeking independent evaluations and even visiting sites where the vendor's ERP is in active use.

The vendor you ultimately choose will have a profound impact on the overall success of your ERP implementation. It is not just a purchase. It is a strategic partnership that will drive your organization's ability to achieve an integrated, efficient and competitive operational framework.

Embrace this process with the same attention to detail and commitment to excellence that defined your exploration thus far. Recognize that while every vendor may offer a solution, perhaps only one will truly partner with you to transform vision into reality, paving the way for long-term success.

As you consider these criteria, remember that the path to ERP success is built on informed decisions, transparent evaluations and a collaborative spirit that bridges the gap between technology and business.

The thoughtful selection of your ERP vendor is not just another step in the project. It is the fulcrum upon which your organization's future operations will pivot. Approach it with intentionality and let this be the beginning of a relationship that drives sustainable, measurable value for your enterprise.

These are the core concepts of the vendor selection process in the ERP execution phase. Use this detailed analysis to understand the importance of aligning vendor capabilities with your business requirements. Understand the need for a comprehensive evaluation of technology, cost, support, culture and the value of viewing the selection process as a strategic partnership rather than a simple transaction.

The shared examples—from legacy data integration challenges to the benefits of agile implementation frameworks—highlight practical steps and real-world lessons that can help you find a vendor who is truly the right match for your organization.

As you proceed, take the time to engage with multiple vendors, ask tough questions and validate their claims with references and detailed demonstrations. This deliberate and measured approach helps you know that your chosen partner is well-equipped to support your transformation from concept to a fully-integrated operational powerhouse.

The right vendor will drive the technical integration and foster an environment of continuous improvement, ensuring that your ERP system remains a vital component of your long-term strategy.

The following are the key characteristics of an ERP vendor that can help decide whether an ERP vendor is the right ERP vendor for your organization.

Committed to Support

Vendor's commitment to provide support and the post-implementation relationship are critical components of the overall evaluation. ERP implementations are rarely static. They evolve as your business grows and as technology advances. It is imperative to select a vendor who is committed to supporting your organization over the long haul.

This means providing excellent technical support during the implementation and offering continuous training, frequent system updates and a partnership approach that evolves over time.

I once observed a vendor that had a comprehensive after-sales support program that included regular system health checks, user refresher training sessions and a dedicated account manager. The client, a financial services firm, found that this level of engagement greatly accelerated the assimilation of the new system into their daily operations.

In contrast, vendors who offer only basic, reactive support tend to leave organizations struggling to bridge the gap between installation and optimal performance.

Committed to Invest in Innovation

Another aspect of vendor support is the responsiveness of their innovation pipeline. The ERP landscape is rapidly evolving, and companies need vendors who are forward-thinking and continuously investing in research and development.

Look for vendors who demonstrate a commitment to innovation by attending industry conferences, releasing regular updates and even incorporating customer feedback into their product roadmap.

On one occasion, a well-regarded vendor announced a roadmap update that incorporated advanced analytics and AI-driven forecasting—a feature that perfectly aligned with a client's strategic goals. This proactive approach to product development reassured the organization that the vendor's product would remain competitive and beneficial well into the future.

Committed to a Collaborative Partnership

It is critical to approach the vendor selection process as a collaborative partnership rather than a one-sided procurement. A vendor who is willing to invest time in understanding your business, ask detailed questions about your operations and tailor their demonstrations accordingly shows that they are serious about being a long-term partner.

In one early-phase meeting with a prospective vendor, I observed how a representative asked insightful questions about the organization's unique process flows and even suggested process adjustments that could yield additional efficiencies. This level of engagement impressed the senior management team and helped forge a sense of partnership that would be invaluable during later phases of the implementation.

Cultural Fit

Beyond the metrics and technical capabilities, pay attention to the cultural fit between your organization and the vendor. ERP systems are as much about relationship management as they are about technology.

A vendor's culture—its approach to collaboration, responsiveness to client needs and overall communication style—should resonate with your own organizational values. If a vendor appears rigid or solely profit-driven, it may be a sign that long-term support will be lacking.

I recall one vendor's approach where every interaction felt consultative rather than transactional. Their flexibility during contract negotiations, coupled with their detailed understanding of the industry challenges, gave a level of comfort that many organizations seek when embarking on an ERP transformation.

Implementation Methodologies

During the vendor evaluation process, it's important to consider the implementation methodologies they advocate. Look for a vendor whose approach to project management, risk mitigation and change management aligns with the strategies you have developed through your thorough planning and review. An implementation methodology that includes structured milestones, regular feedback loops and provisions for scalability and customization is key.

In one notable case, a vendor's agile implementation framework allowed a manufacturing company to move from pilot testing to full-scale deployment in phased increments. This method reduced risks and generated early wins that built momentum and fostered greater organizational commitment to the new system.

At times, a vendor's implementation methodology is not more than a sales brochure to augment their sales process. Dig deeper and ask for supportive data from their prior implementations to contrast the difference between what is being stated and what is being followed on the ground.

The success of your ERP implementation will mostly hinge upon the vendor leadership's representative owning your project along with you. Meet with that person and assess how that person aims to mitigate the risks and resolve implementation challenges highlighted earlier in this book. If you do not feel confident, then clarify your concerns and attempt to establish shared resolution strategies.

Total Cost of Ownership

Cost is another critical factor in the vendor selection process, but it should be considered holistically. While initial costs and licensing fees are important, they represent only a part of the financial commitment. Look at the total cost of ownership, including implementation services, training programs, ongoing support and future upgrade expenses.

In one notable instance, I advised a client who initially selected a vendor based purely on a low upfront cost. However, a thorough analysis revealed that total costs over a five-year period would be significantly higher due to extensive customization needs and poor vendor support. Adjusting for these hidden costs is essential—after all, an ERP system is not a one-time purchase; it is a long-term investment that must deliver sustainable benefits over its lifecycle.

With these insights in mind, you are now better equipped to manage the vendor selection process—a critical gateway towards the eventual execution phase of your ERP initiative. The lessons learned here, combined with rigorous due diligence and a clear focus on strategic alignment, lay the foundation for a partnership that will help turn meticulously crafted plans into a working, dynamic ERP system that serves your organization's true needs.

The vendor selection process, though complex and demanding, is an opportunity to build a foundation of trust, technical excellence and strategic alignment that will propel your organization forward in the age of digital transformation.

Gap Analysis

Gap analysis is a structured approach to identifying discrepancies between current business processes and ERP system capabilities, ensuring the software meets operational needs. The outcome of this process determines whether the gap can be addressed through configuration, customization or process changes.

Following these key steps after your process mapping uncovered gaps that will influence successful ERP implementation.

1. Identify Gaps

- Compare current processes with ERP functionalities to find discrepancies.
- Categorize gaps into "Must-Have," "Should-Have" and "Nice-to-Have"
- Determine whether gaps require customization, process reengineering or third-party integrations.

2. Prioritize Gaps

- Focus on gaps that have the highest impact on business operations.
- Assess feasibility, cost and risk associated with addressing each gap.
- Develop a roadmap for resolving gaps in a phased manner.

3. Develop an Action Plan

- Decide whether to customize the ERP, modify business processes or integrate third-party solutions.
- Establish timelines, responsibilities and success metrics.

Customization vs. Standardization: Finding the Balance

Balancing customization against standardization is a challenge that often defines the long-term success of an ERP implementation.

Too much customization can lead to spiraling costs, complex maintenance issues and difficulties in applying future upgrades, while too little may force your organization to mold its operations into a rigid tool that might not capture your unique process nuances. The art lies in finding a middle ground where the system supports industry best practices and standard functions yet is flexible enough to address your specific operational challenges.

I remember a situation with a mid-sized distribution company that was set on the path of heavily customizing its ERP solution. The executive team believed that every inch of their unique workflow needed to be reflected in the system, and they pressed for custom developments on nearly every module. Initially, the promise of a tailor-made solution was alluring—after all, the new system was supposed to mirror the way they had successfully run their business for years.

However, as the project unfolded, it became apparent that each additional customization added complexity. Every change rippled throughout the system, making future updates a constant headache for their IT department.

This experience taught them that a fully customized solution can sometimes do more harm than good by locking the organization into expensive modifications and increasing the risk of unforeseen glitches. Ultimately, they recalibrated their approach, choosing to keep core functionalities standard while tailoring only key parts where the benefits clearly outweighed the risks.

The decision-making process regarding customization versus standardization often begins with creating a detailed analysis of current business processes. It is essential to ask what aspects of your workflow are truly unique and add measurable value if they are carried forward into the new system.

In one instance, I worked closely with a manufacturing client whose production scheduling was handled by a set of highly specialized algorithms built over years of development. Initially, the temptation was to recreate these algorithms within the ERP environment. However, a thorough cost–benefit analysis revealed that these specialized processes constituted only a small fraction of the overall operations.

By shifting focus to a standardized scheduling module offered by the vendor, the company found that it reduced implementation time and resulted in a scheduling process that was easier to maintain and upgrade. The standardized module still met their needs, with only minor adjustments required to preserve critical features. This outcome saved resources and laid the groundwork for quicker adoption and smoother future upgrades.

In practice, the first step is to engage the key stakeholders from each department who are intimately familiar with the daily nuances of the operations. They understand which parts of the process are truly critical and which ones can be adjusted or improved through standard practices. Workshops that bring together representatives from production, finance, sales and customer support can yield surprising insights.

For example, an automotive parts supplier once discovered that its customized reporting function, which had been developed over several years, was rarely used by the end-users. Instead, they found that relying on the core, standardized reporting capabilities of the ERP system was sufficient, more reliable and easier to interpret across the different departments.

This collaborative review allowed the company to narrow its customization efforts to only those areas where the gains were significant and measurable.

Another significant aspect of this balance is considering the long-term maintenance and upgrade cycles. Standard features tend to come with improvements from the vendor as technology evolves.

For instance, if your system relies heavily on custom code, every time the vendor rolls out an update, your customizations might need to be reworked or re-integrated. This could delay the upgrade process and lead to prolonged periods of instability.

I recall one project where a client faced months of setbacks because the ERP vendor's updates repeatedly conflicted with the extensive customizations they had implemented. In contrast, a company that embraced a mostly standard system enjoyed smoother upgrade cycles and less downtime, as it benefited from the vendor's ongoing commitment to enhancing the core functionality.

The lesson is that while customization can create a competitive edge in the short term, its long-term impact should be carefully weighed against the predictability and upgradeability of standard system offerings.

Fostering a mindset of continuous improvement plays a major role here. Instead of trying to achieve perfection with custom configurations from day one, consider adopting a lean approach by implementing standard functionalities, monitoring performance and only then make targeted adjustments as necessary.

The concept of "minimally viable customization" emerged from several projects I worked on. For instance, one healthcare provider initially opted for a standardized patient data management module. After rolling it out and observing usage patterns, they found a few gaps that could be fine-tuned without a complete overhaul. By adding only minor tweaks—such as creating a custom field for a particularly critical piece of patient information—they managed to optimize the module without undermining the system's long-term maintainability. This approach allowed them to capture the benefits of customization without shouldering the full burden of a highly bespoke solution.

The process of deciding where to customize and where to use off-the-shelf features should involve a rigorous evaluation of the total cost of ownership. Every decision should include the initial development cost, and the ongoing expenses related to training, maintenance and future upgrades. When discussing numbers with the finance team or project sponsors, be prepared with detailed scenarios that compare the long-term benefits of standard versus customized modules.

I've seen several organizations that initially leaned toward heavy customization only to find that the total cost—when factoring in delayed upgrades, the need for specialized support and reduced flexibility—was prohibitive in the long run. A transparent presentation of these comparisons can help align everyone's expectations and reinforce the importance of a balanced approach.

One anecdote that stands out involves a manufacturing company that had aggressively pursued customization to accommodate its unique promotional scheduling. After several rounds of modifications and error-prone deployments, the practical challenges of maintaining the custom solution became evident. Faced with mounting issues and disgruntled staff, the management decided to re-evaluate their approach, opting instead for a standardized promotional module supplemented by periodic customization only where necessary. This pivot streamlined operations and reduced errors and restored confidence among the employees. They quickly noticed that the system was easier to use and that promotions were managed more smoothly, leading to more satisfied customers and a more agile marketing strategy.

It is also imperative to consider how customization affects the scalability of your ERP system. Business needs evolve, and what seems like a perfect solution today may become a bottleneck tomorrow if it is too rigidly tailored.

Standardized solutions offer a degree of predictability and flexibility that allows for modifications without significant system disruption.

A well-known electronics manufacturer once struggled with a highly-customized supply chain module that was optimized for their current operations. However, when the company entered a new market and demand surged unexpectedly, the customized module failed to scale, resulting in supply chain disruptions and lost business opportunities.

Their eventual shift toward a more standardized approach—coupled with minor, targeted customizations—enabled the system to handle increased volumes without a hitch, illustrating the long-term advantages of a balanced strategy.

Moreover, the decision between customization and standardization should be closely linked to your overall change management strategy. If you opt for heavy customization, be prepared for a longer implementation schedule with multiple rounds of testing, fixing and re-testing.

In contrast, a system that relies primarily on standard features typically benefits from a more straightforward series testing, which is consistent with industry norms and more readily supported by the vendor's resources.

For example, a mid-sized services company discovered that its employees quickly adapted to a standard CRM module because it mirrored the interfaces of other common business tools. The ease of adoption translated into faster productivity gains and into lower training costs and fewer support calls during the critical early days of implementation.

It is worthwhile considering the broader implications of customization on vendor relationships. A vendor is more likely to invest in continuous improvements in their core product rather than heavily customized versions that only a few clients utilize. This means that when you opt for a more standardized solution, you benefit from the vendor's ongoing enhancements, new feature rollouts and reliability improvements over time.

There is a risk that by investing in extensive custom solutions, you might isolate your system from forthcoming innovations. In several recent engagements, companies that embraced standard functionalities found that their system evolved more gracefully, absorbing new technological trends without the need for disruptive overhauls.

An effective strategy is to start with a system that adheres as closely as possible to industry-standard practices and then measure its performance relative to your unique metrics. This evaluation should involve pilot tests, feedback sessions with end-users and close monitoring of key performance indicators.

The great news is that most of the modern ERP now offers low-code / no-code enhancement features and those are native to the out-of-the-box software. There is no need to dilute the business process that supports your organization's competitive advantage.

In early 2000, I was engaged to evaluate and recommend an enterprise-wide solution for lean manufacturing. The ERP they had then did not include the functionalities for lean manufacturing. I recommended a custom solution and crafted it. I assembled the team to get that solution developed and implemented globally. That custom solution saved many man-hours and boosted operational efficiency across plants and distribution centers globally. Since that was developed using the native development toolset of the ERP, the future upgrades were uneventful. That was a win-win solution for the stakeholders of that enterprise.

By comparing baseline performance with the outcomes after standard implementation, you can identify areas where targeted customization might yield significant improvements.

For example, an international logistics provider set up a pilot project with their new ERP system, using a standardized warehouse management module. Following a detailed review of performance data, they discovered that a minor custom enhancement in their inventory tracking process could reduce processing times by nearly 20%. The willingness to add only limited, impactful customizations allowed the system to remain robust and upgradable while delivering measurable gains.

Beyond the technical and financial considerations, the cultural dimension within your organization must be addressed. When employees are involved in the decision-making process regarding customization, they are more likely to stick with the chosen solution and embrace the changes it brings. In one project, department heads from several divisions collaborated closely to identify which features were truly essential and which could be streamlined.

Their collective input ensured that the system would meet daily operational needs and fostered a sense of ownership over the new tool. This collaborative approach often results in a smoother transition, as staff feel that the new system has been crafted to address real concerns rather than being imposed from above.

In essence, the journey to strike the right balance between customization and standardization is very relative. It involves scrutiny of your business processes, open collaboration between technical experts and operational stakeholders and a continual assessment of long-term impacts versus short-term efficiencies. The right balance will depend on your organization's specific circumstances, strategic goals and appetite for risk.

It is essential to remain flexible. As your organization grows and your business environment changes, the ideal amount of customization might change. Regular reviews and performance measurements will enable your system to evolve alongside your organizational needs.

The lessons learned from previous ERP projects consistently emphasize that over-customization can lead to future headaches, while a standardized approach offers consistency, easier upgrades and often better support from vendors. However, purchasing a completely off-the-shelf solution may leave your organization with a system that struggles to capture your unique operational requirements.

The art of finding a middle ground is critical. By actively engaging in pilot testing, soliciting user feedback and carefully analyzing performance metrics, you can make informed decisions about where customization is justified and where standardization is the better route.

In the end, your ERP system is not just a software deployment, it becomes the backbone of your daily business operations and a key driver of strategic growth. With the right balance, the system can evolve with technological advances, support continuous improvement and assist every branch of your organization in achieving higher levels of productivity.

The strategic decision to limit customizations to only those areas where clear benefits prevail is one that ultimately pays dividends in operational stability, ease of maintenance and long-term viability.

As you reflect on this exploration of customization versus standardization, consider that the goal is to create an ERP solution that is resilient, maintainable and aligned with your strategic vision. Every modification, every tailored component, must be evaluated on the immediate improvement it promises and on its long-term impact on the system's evolution.

The experience gathered from various industries demonstrates that the most successful implementations are those that embrace a measured approach—one that values simplicity, adheres to standardized practices where possible and only accepts custom solutions when they deliver unmistakable value.

Ultimately, the journey toward a balanced ERP implementation is a journey of continuous learning and adaptive strategy. It requires a firm commitment to thoughtful evaluation at every stage and a willingness to adjust course, based on real-world performance data. As internal processes change and market conditions evolve, the need for recalibration will persist. Your system should be dynamic enough to incorporate these changes without becoming overly complex or burdensome.

As you move forward with your ERP implementation, keep in mind that striking the right balance is about creating a system that is user-friendly and one that integrates seamlessly with your legacy systems, yet is flexible enough to accommodate future changes. In doing so, you may gain confidence knowing that your ERP implementation is not just a one-off project but a transformative tool that drives long-term operational excellence and strategic growth.

With these insights, you are now equipped to make informed decisions about where to invest your customization efforts and where to use standardized solutions. The judicious combination of the two will help you build a resilient, adaptable ERP system that supports your organization's mission and sets the stage for future success. Embrace the process, engage your teams in honest evaluations and maintain an iterative approach. Each adjustment here brings you one step closer to a truly integrated business solution.

Every critical decision regarding customization should be validated by its impact on overall performance, ease of maintenance, the competitive advantages that foster and alignment with strategic goals. By consciously choosing only those overrides that deliver tangible benefits, you reduce the risk of future technical complications and know that your system remains agile, scalable and cost-effective.

The balance between customization and standardization is a pivotal factor in turning your ERP vision into a sustainable reality. The process requires you to understand complex decision trees, involve key stakeholders in meaningful dialogues and continually reassess the cost-benefit equation of each technical tweak.

The experiences shared here—from the pitfalls of over-customization that led to costly upgrades to the benefits of retaining standard functionalities for ease of use and integration—serve as a guide to help you achieve balance.

Let this exploration serve as a benchmark for current decisions and a reminder that the perfect system is not one that eliminates all customization, but one that thoughtfully integrates tailored solutions within a stable, standardized framework. With a clear understanding of the trade-offs involved, you are better prepared to create an ERP environment that meets the demands of today's operations but is also poised for the future.

Now, with the principles of balanced implementation firmly in mind, you are ready to progress to the next phase of the execution process, one where practical steps and tactical adjustments will transform your ERP solution from a well-planned strategy to a fully integrated operational reality.

Process Configuration in ERP

Process configuration in ERP involves mapping and enabling business workflows from the defined future state within the ERP system to ensure efficiency, compliance and scalability. It ensures that the ERP solution supports business objectives without unnecessary complexity.

This step ensures that business workflows are optimized and aligned with the system's capabilities. Here are the key aspects.

1. Workflow Configuration

- Configure ERP modules to fit business processes.
- Define approval hierarchies, automation rules and task sequences.

2. Data Integration

- Ensure seamless data flow across departments.
- Standardize data formats for consistency and accuracy.

3. Automation and Standardization

- Reduce manual tasks by automating repetitive processes.
- Enforce best practices to improve efficiency.

4. User Roles and Permissions

- Define access controls for security and compliance.
- Assign roles based on job functions to prevent unauthorized actions.

5. Compliance and Reporting

- Align ERP processes with regulatory requirements.
- Generate reports for audits, financial tracking and performance analysis.

Custom Development for Gaps

Custom development in ERP involves tailoring the system to meet specific business needs that the selected ERP solution does not fully address. This is a way to resolve the identified gaps. Custom development involves altering core processing logic, workflows, features, adding new applications/modules, integrating with other systems, modules, reports, analytics and insights, etc.

Customizing an ERP can be a complex process depending on the intended objectives. A diligent completion of the following steps delivers outcomes consistently.

1. Create a comprehensive requirements specification document with a clear and specific end goal using a sample dataset, layout, process rules, performance benchmarks and user interface.

2. Develop an exhaustive list of test/use case scenarios with clearly laid out steps, required inputs and expected outputs at every step.

3. Perform a successful completion of unit test, integration test, acceptance test and stress test.

Data Migration: Cleaning and Integrating Legacy Data

Data migration is another critical component that often dictates the success or failure of ERP implementation. Moving data from legacy systems to a new, integrated platform is rarely straightforward. The quality and consistency of your data directly have an impact on the system's overall performance and the accuracy of the subsequent analytics.

I worked with clients who faced the daunting task of cleaning and reconciling decades of data stored in outdated formats. In one memorable project, a manufacturing firm had critical production and financial data spread across multiple systems, each with its own standards and levels of accuracy. The team had to undertake a rigorous process of data cleaning, mapping and integration.

The insight there was clear, which is, a robust data migration strategy is not just about transferring data but about transforming that data into a relatable format that drives more intelligent decision-making. The investment of time and attention in data quality pays off in fewer errors, greater confidence in operational reports and more efficient system performance once the new ERP is live.

Data migration involves extracting, transforming, and loading (ETL) existing data from legacy systems into the new ERP platform while maintaining data integrity, accuracy and consistency.

Key Phases of Data Migration

Here are the five key phases of the data migration process.

1. **Data Assessment and Cleansing**
    - Identify the data to be migrated, including financial records, customer details, inventory data, etc.
    - Cleanse and standardize data to remove duplicates, inconsistencies and obsolete information.
    - Ensure data conforms to the formats and requirements of the new ERP system.

2. **Data Mapping and Transformation**
    - Map data fields from the legacy system to corresponding fields in the ERP solution.
    - Convert data formats to align with the new system's architecture and rules.
3. **Data Migration Strategy Selection**
    - "Big Bang" migration: Moving all data in one go—ideal for smaller datasets but risky for large enterprises.
    - Incremental migration: Transferring data in phases to reduce downtime and mitigate risks.
4. **Testing and Validation**
    - Conduct rigorous testing to identify errors and inconsistencies in the migrated data.
    - Validate data accuracy by running reconciliation reports and comparing legacy system results.
5. **Go-Live and Post-Migration Optimization**
    - Use a "cut-over" data migration strategy for a smooth transition.
    - Monitor system performance and data integrity post-deployment.
    - Train employees on the new ERP workflows to ensure data is properly utilized.

Key Considerations for Business Leaders

Here are the key considerations for business leaders.

- **Data Governance:** Establish clear policies for data ownership, access control and compliance.
- **Stakeholder Collaboration:** Involve IT teams, department heads and ERP vendors to ensure alignment.
- **Risk Management:** Identify potential migration pitfalls and have contingency plans.
- **Business Continuity:** Ensure operations are minimally disrupted during the transition.

Best Practices for Data Governance

Effective data governance ensures that data remains accurate, secure and compliant while supporting business objectives.

Here are the best practices:

Establish Clear Data Ownership and Responsibilities

- Assign data owners to oversee data quality and compliance.
- Define roles and responsibilities for who can access, modify and audit data.

Develop and Enforce Data Policies

- Create policies for data storage, access control and retention.
- Implement data classification (e.g., confidential, public, restricted).

Ensure Data Quality and Integrity

- Use automated validation checks to detect errors and inconsistencies.
- Regularly conduct data audits to maintain accuracy.

Implement Strong Security Measures

- Encrypt sensitive data to protect against breaches.
- Use multi-factor authentication (MFA) and access controls.

Enable Compliance and Regulatory Adherence

- Align data governance with GDPR, CCPA, HIPAA or other relevant regulations.
- Maintain audit trails for data access and modifications.

Standardize Data Across the Organization

- Define consistent naming conventions and formats for data entry.
- Ensure data standardization across all departments.

Invest in Data Governance Technology

- Utilize data governance platforms for automation and monitoring.
- Implement metadata management to track data lineage and usage.

Encourage a Data-Driven Culture

- Train employees to follow data governance policies and best practices.
- Foster collaboration across departments to ensure data consistency.

ERP systems thrive when there is a single source of truth—consistent data definitions, uniform coding standards and authorized user access protocols. As you incorporate more sophisticated analytics, complexities can quickly arise if system foundations are poorly executed or if data quality is not rigorously maintained.

Consider a multinational manufacturer that initially hoped to apply AI-driven demand forecasting. They soon discovered that their master data, although centralized, contained inconsistent item codes and incomplete records inherited from legacy systems. Such inconsistencies undermine even the most advanced analytical models.

The company had to invest significant effort in data cleansing, standardization and re-mapping processes. These improvements paid off. Once data quality was assured, the predictive modules within their ERP produced far more accurate insights, reducing forecast errors and enabling more precise resource allocation.

The lesson here is that advanced analytics in an ERP context rest on solid data hygiene and well-defined governance practices.

Testing: Did We Get It Right?

Testing the implementation of an ERP system is a crucial step to ensure the software operates correctly, integrates well with existing processes and delivers the expected value to the organization. The testing phase typically begins with a clear test plan that outlines the scope, objectives, roles, responsibilities, timelines and testing tools. This plan should align with the ERP implementation goals and include both functional and non-functional testing to verify that all system requirements are met.

A robust testing strategy in ERP implementation ensures a seamless transition while minimizing risks and disruptions. Business leaders must prioritize thorough testing to validate the system's functionality, data accuracy and integration with existing processes.

Key Phases of an ERP Testing Strategy

Here are the eight phases of a comprehensive ERP testing strategy.

1. **Unit Testing**
    - Validate individual modules or components of the ERP system.
    - Ensure each function works correctly before integration.
2. **Integration Testing**
    - Assess interactions between different ERP modules (e.g., finance and supply chain).
    - Ensure seamless data flow and compatibility across systems.
3. **System Testing**
    - Conduct end-to-end testing to verify overall system performance.
    - Simulate real-world scenarios to assess stability and responsiveness.
4. **User Acceptance Testing**
    - Involve key stakeholders and end-users in testing.
    - Confirm system usability, workflows and alignment with business needs.
    - Gather feedback for final refinements.
5. **Performance and Load Testing**
    - Test system speed and responsiveness under varying workloads.
    - Identify potential bottlenecks and optimize system capacity.
6. **Security Testing**
    - Validate data security, access controls and compliance measures.
    - Ensure protection against vulnerabilities and cyber threats.
7. **Regression Testing**
    - Re-test modified modules to confirm changes haven't affected existing functions.
    - Maintain system consistency across updates.

8. **Go-Live Simulation**

   - Run trial deployments in a controlled environment.
   - Verify readiness before full-scale implementation.

Best Practices for Testing

Here are the industry-wide best practices for ERP testing.

- **Define Clear Testing Objectives:** Align testing with business goals and key performance indicators.

- **Engage Cross-Functional Teams:** Involve IT, finance, HR and other departments for a comprehensive evaluation.

- **Utilize Automation Tools:** Enhance efficiency with automated testing for repetitive tasks.

- **Monitor and Document Issues:** Maintain a structured approach to tracking bugs and resolutions.

- **Prioritize Data Validation:** Data is at the core of any transaction and hence ensures that the master data is pristine and appropriate to business operations.

Common Obstacles in Testing

Organizations often face significant challenges that can impact system deployment. Here are some of the most common obstacles business leaders should anticipate.

Incomplete Test Scenarios

- Organizations may overlook real-world business processes during testing.
- Lack of comprehensive test cases can lead to operational disruptions post-launch.

Data Migration Issues

- Inconsistent or inaccurate data from legacy systems can cause failures.

- Data formatting mismatches can lead to errors in financial reporting and inventory management.

Insufficient User Participation

- If end-users and key stakeholders are not actively involved, critical usability issues might be missed.
- Employees may struggle with the new system due to a lack of training.

System Integration Complexity

- ERP systems must seamlessly interact with third-party applications like CRM, payroll and procurement platforms.
- Integration errors can lead to data inconsistencies, transaction failures or workflow disruptions.

Performance Bottlenecks

- Organizations may face slow response times due to poor load testing.
- High volumes of concurrent transactions can overwhelm system capacity.

Security Vulnerabilities

- Weak access control and authentication mechanisms may expose sensitive data.
- Failure to conduct penetration testing can lead to security breaches post-deployment.

Resistance to Change

- Employees and management may be reluctant to adopt the new ERP system.
- Poor communication about system benefits and improvements can hinder acceptance.

Budget and Time Constraints

- Testing often requires significant resources, but organizations may limit funding.

- Rushed implementation due to tight deadlines can lead to untested functionalities.

Lessons Learned During the Execution Phase

This section summarizes the key themes and practical insights that guided us from the drawing board to a working, integrated system. It is also meant to prepare you for the next stage, where the focus shifts to harnessing the full potential of your new ERP system for long-term operational excellence.

The selection of an appropriate vendor and platform is a decision that lays the foundation for everything that follows. Conducting a disciplined evaluation based on a detailed requirements document, comprehensive cost analysis and an understanding of both the technical and cultural fit is essential.

I recall the story of the mid-sized company that initially partnered with a vendor simply due to strong executive relationships. When the implementation revealed that the system only met a fraction of their operational needs, it became clear that relationships and flashy promises could not replace due diligence. Instead, the need for independent assessments and reference visits emerged as a guiding principle. The lesson here is that the vendor you choose will be your strategic partner, and ensuring the right fit protects your investment over the long term.

There is a delicate balance between customization and standardization. As the narrative unfolded, it became increasingly apparent that over-customization can lead to spiraling costs, difficulties in maintenance and challenges in future upgrades.

I shared the experience of a distribution company that pursued heavy customizations at every turn. In those early days, tailored solutions appeared to offer an advantage; however, as the project advanced, every added custom feature complicated subsequent upgrades and prolonged downtime during releases.

This taught the important lesson that a balanced approach – one that retains core standard functionalities while allowing for minimal, targeted modifications – often yields the best long-term results. It is not about eliminating customization altogether, but about being judicious in every decision. The key is to align those custom solutions only with certain processes that drive measurable improvements, while leaning on proven, off-the-shelf capabilities for everything else.

Another major focus was on practical execution. At the heart of execution is the matter of making high-level decisions and the daily management of the project. Don't overlook the importance of establishing realistic timelines, setting budgets that account for contingencies and instituting a robust communication framework among all stakeholders.

I shared several anecdotes—from grappling with unexpected data migration challenges to employee resistance—each reflecting that execution is an iterative process. In a notable instance, a manufacturing company implemented daily review meetings during its data integration phase, ensuring that every detected glitch was addressed in a timely manner. This transparency and continuous improvement helped maintain momentum and fostered a sense of collective responsibility.

The bottom line is that execution must be flexible and adaptive. If problems arise, reallocation of resources, adjustments in training approaches or even renegotiation with vendors may become necessary. Embracing this adaptability, while keeping the broader strategic objectives in focus, is what separates successful projects from those that falter under unexpected pressure.

Taken together, these areas underscore a central message, which is, the execution phase is where planning and theory are tested by real-world challenges. It is here that a well-thought-out strategy is validated or must be refined.

Your ability to select an ERP partner with a proven track record, strike a productive balance between customizing key processes and relying on standard functionalities and manage the daily operations with a diligent, transparent approach will dictate the overall success of your project. In each instance, effective project management is an ongoing process. Each setback and necessary adjustment contribute to a learning process that strengthens the outcome.

Consider the example of a mid-sized manufacturing firm that struggled with data integration issues from legacy systems. Initially, the project appeared stalled as the team encountered incompatibilities and delays, threatening to significantly push back the launch date.

However, by revisiting their execution strategy, they methodically conducted a comprehensive data inventory, refined their data mapping processes and adjusted the timeline sufficiently to allow for proper testing. They also established regular check-ins and used these opportunities to reassess the situation.

What might have been a project failure turned into a robust, well-functioning integration. Success lay in their willingness to acknowledge that challenges were inevitable and that resolutions required technical fixes and process improvements. These lessons reiterate that technology, people and processes must evolve together.

It is important to emphasize that the execution phase is not a final destination but part of a continuous process. Each technical glitch, every adjustment in timeline or budget and each session of intensive training contributes to a cycle of constant improvement. The strategic decisions made now lay the foundation for the ERP system to evolve over time, adapting to new market conditions, organizational changes and technological advancements.

This is where the long-term benefits of your initial planning truly begin to manifest. Your execution phase is a living, dynamic process that, when continuously refined, will transform your organization's approach to everything from daily operations to high-level strategic decision-making.

To summarize the key themes of this chapter, remember these essential points:

- The vendor selection process must be objective, thorough and aligned with your operational requirements, so that the chosen solution is capable and scalable.

- Striking a balance between customization and standardization is essential to avoid the pitfalls of excessive tailor-made solutions while capturing the benefits of industry-standard efficiencies.

- Effective execution requires a robust framework, including realistic timelines, well-planned budgets and open communication channels that facilitate continuous improvement and cross-departmental collaboration.

- Technical, human and external challenges are interrelated. Success in overcoming them depends on a flexible, adaptive strategy rooted in proactive problem solving and ongoing risk management.

- The path from planning to practice is iterative, and each setback is an opportunity to refine and optimize the system, ensuring it remains a reliable backbone for your organizational goals.

These lessons should inform your approach as you move forward. Execution is about turning careful strategy into effective operations. It is the critical link between planning and performance.

CHAPTER 6

THE HUMAN FACTOR IN ERP SUCCESS

Chapter 6: The Human Factor in ERP Success

In every ERP implementation, technology and processes form only part of the equation. Over the decades, I learned that the human element—how people adapt, learn and take ownership of new ways of working—is what truly determines whether an ERP system drives lasting success.

With a carefully chosen vendor, a balanced blend of customization and standardization and a well-executed rollout, you might think the most challenging work is done. However, even the best technical solution can fall short if the people it is meant to support are not fully engaged and prepared for change. This chapter centers on that vital aspect—the human factor in ERP success.

From the earliest stages of planning, I observed that the resistance to change, gaps in training and inadequate communication can derail an ERP initiative even if every other component is flawlessly executed.

I recall a project with a manufacturing company that had invested millions in upgrading its systems. Despite having strong technical support and a comprehensive implementation plan, the project struggled in the months following the launch. The issue wasn't the system itself. It was the people.

Front-line staff felt overwhelmed by the new user interface, middle managers were unsure about their revised roles and senior leadership found that the promised improvements in decision-making were not materializing because user data was not being interpreted correctly.

Performance metrics began to improve only when concerted efforts were made to train, support and communicate openly with all levels of the organization. In that instance, the human dimension emerged as the decisive, transformative element.

This chapter is dedicated to exploring how you can build a culture of continuous improvement, drive targeted training and development initiatives and implement communication strategies that foster an environment where change is embraced rather than resisted. Learn from practical methods and real-world experiences that illustrate how supporting users can dramatically enhance the return on your ERP investment.

Building a Culture of Continuous Improvement

The human element of ERP implementation is often described as the heart of transformation because it is the people within an organization who bring the system to life.

Explore how fostering a culture of engagement, continuous learning and open communication is essential for realizing the full potential of an ERP investment. These are the essentials of the human factor, which includes building a resilient change culture, creating effective training and development and aligning communication strategies to support a smooth transition.

Real-world experiences illustrate how these elements converge to turn a technological initiative into a transformative force for the entire organization. In many organizations, the introduction of a new ERP system signals more than just an upgrade in technology—it signals a change in how work is done.

I recall a well-known financial services firm that implemented an ERP system with a strong technological foundation. Despite the impressive technical capabilities, the company initially experienced employee-dependent bottlenecks in areas like report generation and decision-making.

Embrace Change

A key facet of this discussion is the concept of building a culture that does not fear change but actively seeks to improve. When an ERP system is implemented, it represents not just a new software platform but a new way of thinking about business operations. The culture shift required is substantial. For every successful ERP project, a deliberate strategy was put in place to transform skepticism into enthusiasm.

In one memorable example, a medium-sized manufacturing firm revamped its internal training program just weeks before the launch. The company organized interactive workshops, hands-on training sessions and created open forums where employees could ask questions without fear of criticism.

By involving team members in every step—from the initial design of training materials to the final demonstration of system capabilities—they created a sense of ownership among staff, and the transformation took on a life of its own. What might have been a simple system rollout became a collaborative process of learning and advancement.

Alignment with Daily Tasks and Workflows

Moreover, aligning the ERP system with your organization's daily practices and natural workflows is essential. Too often, systems are implemented without a proper assessment of how people work each day, leading to processes that appear cumbersome or detached from reality.

In one instance, a supplier of industrial components attempted to force its employees to adopt a new workflow that, while technically efficient, did not take into account the on-the-ground challenges faced by its staff. The result was frustration, reduced productivity and a notable drop in morale.

After serious discussions and user feedback sessions, the company refined the workflow to align more closely with the natural cadence of daily work. This adjustment improved operational metrics and reinforced the belief that the change was designed to aid rather than hinder the workforce.

External Dimensions of Change Management

Change management extends beyond the walls of your organization to involve clients, partners and even vendor relationships. If your employees are equipped to embrace change, they become better ambassadors for these external stakeholders.

For example, a company I worked with in the healthcare sector used its ERP system as a platform for internal improvement and for enhancing patient service. By using extensive staff training in technical and interpersonal communication skills, the firm was able to integrate patient feedback directly into system updates, leading to better service outcomes and a more responsive operational model.

This holistic approach underscores that the human factor touches every part of the ERP ecosystem.

The Integral Role of Leadership

Throughout these explorations, the topic of leadership emerges repeatedly. Leaders play an integral role in modeling the behavior that drives a successful ERP implementation. A strong leadership team sets the tone for the entire organization. From proactively addressing concerns to celebrating small victories early in the implementation, effective leaders build momentum.

A memorable example comes from a multinational corporation where the CEO personally visited various departments to observe the new system in action. This visible commitment from the top echelon of leadership instilled confidence among employees and signaled that the ERP transformation was a key strategic priority.

When leadership takes an active role, the human element of the project is strengthened, with employees more likely to embrace the system as a tool for personal and corporate success.

When the human aspect is given the same level of importance as the technological components of your ERP system, the result is a strong, adaptable organization that is primed for future growth. It's about transforming the work culture so that every employee, at every level, perceives the ERP system as an indispensable resource and a means to enhance the efficiency of their work and contribute to the overall success of the company.

Establishing a culture of continuous improvement and fostering an environment where supported users drive advancement are not final steps but ongoing endeavors. The strategies discussed here—splitting training into phases, maintaining open communication channels and involving employees in decision-making—are designed to build a resilient foundation for your ERP implementation. Through these efforts, your organization adapts to the new system and uses it as a platform for continual progress.

Whether you're leading a small business or managing a global enterprise, the core challenge remains the same, which is, making sure that the people in your organization are comfortable with change and are enthusiastic participants in driving that change forward.

Every organization has its own culture, its own challenges and its own strengths—and by aligning these with a carefully constructed human resource strategy, you create a platform upon which the technological benefits of an ERP system can shine.

As you engage with the methods and narratives in this chapter, consider how you might apply these insights within your own organization. What are the current barriers to change within your teams? How can leadership better communicate the value of the new system? In what ways can training and continuous improvement be structured to foster rapid adoption and sustained engagement? Addressing these questions is not a one-time exercise but an ongoing conversation that will evolve with your ERP implementation.

The success of your ERP initiative depends on the strength of your human capital. Continuous improvement, comprehensive training programs and strategic communication support the initial implementation and drive long-term operational success.

By focusing on supporting your employees, you can be certain that the new system does more than streamline data and automate processes. It also transforms daily operations and supports the entire organization.

Changing Routines for Improvements

Fundamentally, successful ERP adoption requires a culture that embraces change rather than fears it. This means moving away from the notion that "this is how we have always done it" and instead developing a mindset that sees change as an opportunity for growth.

In one case, resistance to change was substantial. A deep examination revealed that the root cause was not the system's design; rather, it was unfamiliarity and discomfort among the staff with a tool that radically changed their routines.

What mattered most was not the code, but the readiness of employees to adopt a new way of working. Leaders at the firm eventually recognized that the ERP system would only deliver on its promises if the staff were adequately trained and continuously supported.

They established a series of cross-departmental training sessions and feedback forums that allowed employees to learn at their own pace and voice their concerns openly. When the human element was prioritized, the system's technical strengths came to the fore and productivity improvements followed.

Or consider a manufacturing company that relied on legacy spreadsheets and disconnected systems for decades. For many employees, those tools were practical and familiar. When the new ERP system was introduced, initial resistance was high because the staff felt that their well-honed processes were being radically altered.

In response, management launched initiatives designed to bridge the gap between the old and the new. They organized workshops where employees could see how the new system simplified tasks that used to be laborious. Trainers worked side-by-side with staff, customizing training sessions to address specific pain points and ensuring that every question was answered.

Over time, the workforce began to appreciate that the change was not about discarding their expertise but about augmenting it with a tool capable of transforming operational efficiency. This transformative process required technical sessions and emotional and motivational support from leadership, a reminder that change management is as much about human inspiration as it is about systems.

Alignment with Existing Cultural Frameworks

A recurring theme in successful ERP implementations is the need to align the new system with the existing cultural frameworks of the organization. In many cases, the most productive employees are those who see the ERP system not as an imposition but as an enhancer of their work life.

For example, a healthcare provider that recently implemented a new system found that its success depended on how well the new workflows meshed with the organization's patient-centric values. Rather than forcing the existing processes to change overnight, the implementation team worked closely with department leaders to integrate familiar elements into the new workflows.

The result was a system that felt intuitive, even as it introduced modern efficiencies. This approach highlights that successful change management is not about radical disruption but about evolving practices gradually, ensuring that every member of the organization feels neither overwhelmed nor alienated.

Incorporate Feedback

An essential factor in mobilizing human participation is incorporating feedback mechanisms across all levels of the organization. Establishing clear channels through which employees can report issues, suggest improvements and express concerns is vital.

One notable instance involved an ERP roll-out at a manufacturing plant where a structured feedback program was created. Employees submitted suggestions via an online portal, and weekly meetings were held to review this feedback. The most effective improvements made to the system emerged directly from these sessions, such as adjustments to interface layouts and modifications to reporting templates that significantly reduced manual data entry.

These practical changes improved operational efficiency and demonstrated to employees that their input was valued, thereby strengthening their commitment to the transition.

The transformation facilitated by effective human resource strategies does not end once the ERP system is fully implemented; instead, it sets the stage for a culture of continuous advancement. When employees see that the new system is a living, evolving platform that adapts to their needs, they are more likely to embrace a mindset of ongoing improvement. This cultural shift can lead to advancements that extend well beyond the initial implementation.

For instance, a multinational logistics provider adopted an ERP system and used it to streamline their route optimization processes. Encouraged by the new system's analytical capabilities and the open culture established during training, staff began experimenting with data-driven improvements.

Over time, these incremental advances resulted in significant cost savings and enhanced service levels. Here, the ERP system served as the catalyst for operational change and as a platform that supported employees in driving further improvements.

The success of an ERP implementation is predicated on nurturing the human element with the same care and attention as technical components. Engaging employees, fostering continuous learning, nurturing open communication and compelling leadership are all essential ingredients for a successful transformation.

When these human-centered strategies are integrated into the overall ERP rollout, the system becomes more than just a technical asset. It becomes a reflection of the organization's evolving culture.

Key Takeaways

Here are some key takeaways to consider:

- Organizations investing in the human factor embrace change, adapt and benefit from ERP systems.

- Initiatives like training, mentorship, leadership engagement and feedback measure readiness.
- Nurturing the human side fosters a cycle of improvement, driving success through technology and people.
- Building understanding encourages commitment, showing ERP as a tool to enhance work, not to replace expertise.
- Prioritizing training, dialogue and leadership prepares organizations for a successful, sustainable ERP implementation.

Training and Learning: User Enablement

Effective training, learning and strong communication strategies form the lifeblood of successful ERP implementations, shaping the way employees connect with new systems and driving the cultural transformation essential for lasting success.

In this section, let's explore in-depth methods and best practices for equipping your workforce with the skills and confidence needed to manage a complex ERP environment. Our discussion goes beyond standard training sessions to address continuous learning, personalized support and the critical role that open dialogue plays in fostering people's trust in change.

The insights shared here come from years of experience and real-world examples that shed light on how training and communication can turn potential resistance into genuine enthusiasm and adoption.

Training is an Evolving Process

One recurring insight is that training must be seen as more than a one-off event, as an evolving process that adapts to the technology and the changing needs of your organization over time. When an ERP system is introduced, employees may feel overwhelmed by the novelty of the interface and by concerns that their expertise in legacy systems will be rendered obsolete.

In one memorable case, a manufacturing firm initially scheduled a series of training sessions immediately before and after the ERP launch date. Yet, within weeks, managers noticed that while the sessions imparted basic knowledge, employees were still unsure how to apply the new system functionalities to real-world scenarios.

Recognizing this gap, the company implemented an ongoing learning program that combined periodic refresher courses, hands-on workshops tailored to daily operations and peer-led mentor groups.

Over time, this continuous approach reduced errors in system usage and encouraged frontline staff to suggest improvements and share tips with colleagues. The manufacturing firm's experience underscores that continuous training builds a resilient culture where learning becomes a shared, ongoing responsibility.

Tailored Training

A key element of enduring training programs is the development of training content that is tailored to specific departments and user roles. Generic sessions that cover only the surface of system functions may leave many users in the dark about how the ERP system directly benefits their daily tasks.

For instance, for a distribution company, I once consulted with a supply chain director, logistics manager and the operations staff who had vastly different interactions with the new ERP platform. Initially, the training sessions treated all users uniformly, focusing on technical functionalities without considering role-specific applications.

Soon, it became evident that further persistence in a generic format led to significant frustration and a steep learning curve among technical and non-technical staff alike. In response, the organization restructured its training program into specialized tracks that addressed the unique needs of each group.

The supply chain director received lessons on streamlined supply chain optimization and real-time reporting. The logistics managers were trained on inventory data integration and route evaluation features, while operations staff learned how to use dashboards and mobile devices to perform their tasks.

By aligning training content with the specific work processes and responsibilities of each department, the institution saw a quicker adoption curve and increased satisfaction and confidence among employees. This example illustrates that an effective training program is rooted in understanding the end-user's perspective and tailoring programs accordingly.

Multi-Modality Training

Beyond the content of the training, the delivery mechanism itself plays a critical role. Traditional classroom-style sessions may work well for a few users, but a blended learning approach that incorporates interactive e-learning modules, simulation exercises and even virtual reality scenarios can enhance understanding and retention.

For example, one manufacturing company adopted an interactive e-learning platform that allowed employees to learn by using simulated ERP system interfaces. This self-paced learning lets users experiment and make mistakes in a risk-free environment, thereby building their competence before handling live data.

This approach increased participation rates and allowed managers to track progress and identify areas where further coaching was needed. In a fast-evolving digital landscape, adaptive learning systems that create real-time feedback and personalization can transform standard training into an immersive experience that resonates with a broad spectrum of learners.

Fostering Continuous Learning

A further aspect of harnessing the human factor is fostering a continuous learning environment. Many companies treat the training phase as a one-time event, something to be checked off during implementation. However, the true strength of an ERP system emerges only when learning is seen as a continuous process.

In one case involving a mid-sized services company, the initial training program was extensive but soon faded into complacency. Recognizing this gap, the leadership instituted a series of refresher sessions, ongoing e-learning modules and peer mentoring programs. This approach reinforced the functionalities of the ERP system and allowed the organization to keep pace with system upgrades and process refinements.

As employees become more proficient, they begin to contribute valuable feedback that further shapes the system. This iterative cycle of learning and improvement transforms the ERP system from a static tool into a dynamic resource that continuously evolves along with the business.

At the heart of this human-centric approach is the idea that training is an ongoing process. How can you be certain that employees who have spent years working in familiar systems are able to use the new ERP system and can leverage it to gain a competitive edge? The answer lies in continuous training and development programs that adapt to the evolving technology and the changing needs of the workforce.

Real-world case studies illustrate that organizations that invest in frontline training often realize a quicker adoption curve, fewer errors and better operational outcomes.

Leadership Participation

The role of leadership in driving a continuous learning culture cannot be overemphasized, either. Leaders must communicate their commitment to the ERP initiative by participating actively in training sessions, attending feedback meetings and regularly reiterating the system's strategic benefits.

It is also important to consider how leadership influences the success of ERP adoption. Leaders are not just decision makers or strategists; they are also role models who guide behavior and set the tone for change. When a CEO takes time to visit departments and observe the system in action, it sends a clear message that the transformation is a strategic priority.

I recall the experience of a global conglomerate where the CEO made it a point to personally lead quarterly review meetings during the ERP transition. His presence, along with a willingness to listen to frontline feedback, played a significant role in building morale.

Leaders who are visible and accessible help to humanize the process, making the change less intimidating. Their commitment to regular, two-way communication, coupled with a demonstrable willingness to act on feedback, builds an organizational culture where advancement and adaptation are part of the daily routine.

After the first month of rollout, I recall a CEO who personally hosted several departmental meetings to discuss early challenges and successes. His openness about the difficulties encountered and his willingness to listen to employee opinions helped dispel initial skepticism.

By framing setbacks as opportunities for growth and continual improvement, he inspired a mindset where challenges were viewed not as failures but as integral components of a long-term transformation. This form of leadership fosters an environment of trust and resilience, where every employee is motivated to contribute toward building an operational framework that can adapt to future demands.

Leadership for Cultural Transition

Leadership plays a vital role in modeling the behavior necessary for a smooth cultural transition. Leaders who personally participated in training sessions, attended feedback meetings and even took time to walk the production floor sent a powerful message that this transformation matters at every level.

In one instance, the CEO of a multinational conglomerate held regular review sessions to discuss the challenges and positive outcomes of the new system. His presence reaffirmed that the move to an integrated ERP system was a strategic priority and demonstrated that overcoming challenges was a collective responsibility.

This kind of visible, accountable leadership helped shift the mindset from “we’re losing what we know” to “we’re gaining an opportunity to work smarter. " It also established an atmosphere where every employee felt connected to and responsible for the success of the implementation.

Moreover, integrating these elements—continuous learning, effective communication and strong leadership—helps create an enduring cultural shift that extends far beyond the ERP project itself.

When employees are continuously engaged and supported, the system evolves from being a static piece of software into a dynamic framework that adapts effortlessly to changing market conditions and internal demands. The initial resistance fades away as individuals grow comfortable with the technology and with the idea of contributing to an ongoing process of improvement.

This cultural evolution is not instantaneous. It requires persistent effort and regular reinforcement through feedback, acknowledgment of successes and commitment to addressing challenges head-on.

One of the most powerful insights in our discussion has been that change is not a momentary disruption; it is an ongoing process. Each step taken, from the first training session to the continuous adaptation of communication protocols, contributes to an organizational culture characterized by resilience and agility.

The examples—from a manufacturing firm that transformed data integration issues into streamlined workflows, to a manufacturing company that turned employee apprehension into a source of advancement—illustrate that the strength of an ERP implementation lies in its human components. When people feel supported, the new system becomes more than just a tool. It becomes the nerve center for decision-making, operational efficiency and strategic growth.

A recurring observation is that training and communication should not be static, isolated events relegated to the beginning of the project. Instead, they must be woven into the organization's everyday life.

In one memorable case, a healthcare provider did not limit its training to the initial rollout, but continuously updated its curriculum as new features and reporting tools were introduced. This adaptable approach created a smooth integration process and helped drive improvements in patient service and operational transparency.

Furthermore, it is essential for senior management to recognize that the success of the ERP initiative depends on making every stakeholder feel like an active participant in the transformation, not just passive recipients of change. When employees are invited to give feedback, when their suggestions are implemented and when successes are celebrated, a collaborative spirit is nurtured.

A mid-sized services company that implemented an online feedback portal saw an increase in user satisfaction and tangible improvements in system performance. Employees actively identified redundant steps that, once eliminated, saved significant processing time. This collaborative atmosphere transformed frustrations into opportunities, with everyday users becoming champions for the system rather than barriers to its full adoption.

Peer-to-Peer Mentoring

Investing in peer-to-peer learning and mentoring is another element that deepens the human factor. Once the ERP system is in place, employees who quickly become proficient can serve as mentors for their colleagues who may be struggling initially. Peer mentoring programs enhance the technical proficiency of the workforce collectively and help to build solidarity and be certain that knowledge is distributed evenly throughout the organization.

For instance, peer mentoring programs and learning labs can create ongoing opportunities for employees to familiarize themselves with system updates and share best practices. Likewise, establishing regular review sessions, where progress is measured, challenges discussed and improvements planned, helps the ERP system evolve continuously in line with business needs.

In one prominent example, a manufacturing firm organized a "buddy system" where experienced users were paired with newcomers. This initiative led to a marked improvement in problem resolution times, as well as an increase in employee confidence and system adoption. The collaborative spirit fostered in these programs can be a lasting asset beyond the implementation of ERP, establishing a culture where continuous learning and mutual support become the norm.

Reward Programs

Effective training and communication strategies are further enhanced by performance metrics and rewards programs that acknowledge progress and celebrate success. Recognizing and rewarding milestones can act as powerful motivators.

For instance, I know of successful organizations that give awards for teams that achieve the highest system proficiency improvements and for departments that consistently create valuable feedback leading to process enhancements. Such initiatives build excitement and reinforce a performance-driven culture where every improvement is valued.

Over time, these rewards contribute significantly to a climate of continuous improvement, where employees strive to meet baseline expectations and exceed them, knowing that their contributions have a direct impact on the organization's overall success.

Integrating Training with Broader Strategic Goals

Another essential aspect is integrating training and communication efforts with broader organizational goals. The ERP system should not be seen as an isolated project but as an integral part of the company's strategy for growth and advancement. When employees understand how the system ties into strategic objectives—such as improving customer service, reducing operational costs or enhancing decision-making—they are more likely to invest emotionally in its success.

In the case of a manufacturing company, management made a concerted effort to link training sessions to essential KPIs such as order fulfillment times and customer satisfaction scores. By demonstrating how system proficiency could directly translate into improved performance metrics, employees became more motivated to engage fully with the training program.

These clear connections between individual efforts, system performance and overall business success help crystallize the ERP initiative's value in their daily work.

Create a Learning Lab

One useful practice that emerged from several successful implementations is the idea of creating a "learning lab" environment—the digital equivalent of a sandbox—where employees can explore system functionalities without fear of affecting live operations.

In one project, the IT team set up a parallel test environment that mimicked the live ERP system as closely as possible. Employees were encouraged to experiment, try out new reports and reconfigure dashboards to suit their needs. This hands-on, exploratory approach fostered a deeper familiarity with the system and led to early identification of potential issues that might have otherwise persisted unnoticed.

In addition, these sessions often sparked creative ideas on how to use the platform more effectively, driven by the insights of daily users. The learning lab became a vibrant hub of advancement, bridging the gap between routine operations and strategic improvement.

State-of-the-System Reviews

An additional strategy for deepening the human impact of the ERP transformation is ongoing cultural reinforcement. Change management experts advise that once an ERP system has been implemented, the work is far from over and that maintaining enthusiasm and adoption requires continual reinforcement of the new culture.

This can be achieved through periodic "state-of-the-system" reviews, where leadership and project teams come together to review progress, celebrate successes and plan for future enhancements.

In one organization, quarterly reviews became an anticipated event where employees could see the tangible benefits of their efforts—ranging from improved workflow efficiencies to financial savings—and hear directly from senior executives about the system's role in driving strategic goals.

This recurring reinforcement helps the ERP system to remain at the forefront of operational considerations and builds a legacy of continuous improvement and advancement.

Harnessing the human factor through effective training and communication is not just about the initial rollout—it is an ongoing commitment to evolving your organization's culture. It means building a foundation where every employee, from the lowest-level staff to the highest echelons of leadership, is engaged, informed and supported to contribute actively to the success of the ERP system.

The shift in mindset—from viewing change as a disruption to seeing it as an opportunity for growth—is what distinguishes organizations that thrive in a fast-paced environment from those that struggle to adapt.

To bring these ideas full circle, let's reiterate key insights:

- Continuous training that evolves with the system is essential. Do not limit learning to initial sessions; instead, develop a cycle of refresher courses, interactive workshops and peer mentoring.

- Communication should be open, frequent and two-way. Establish structured channels and feedback mechanisms to be certain that every employee's voice is heard and that updates are shared transparently.

- Leadership must be visible and actively engaged, modeling the behavior they wish to see and reinforcing the strategic importance of the new ERP system through regular interactions and supportive policies.

- Integrating training with performance metrics and rewards can drive a culture of excellence, linking system usage directly to tangible business outcomes.

- Creating a sandbox or learning lab where employees can safely experiment with the ERP system encourages advancement and real-time problem-solving, making users more comfortable and proficient.

Recognize these insights as a blueprint and a call-to-action to prioritize people throughout the transformation process. Understand that every session of hands-on training, every open forum for feedback and every leadership visit will shape the success of your ERP system and, by extension, your organization's future. It is this dedication to continuous learning and open communication that will transform a complex technical project into a thriving cultural evolution.

As you reflect on these ideas, consider the ways in which ongoing training, effective communication and active leadership have already begun to reshape your organizational culture. Think about the practical steps you can take to be certain that every employee is proficient with the new ERP system and feels personally invested in its success. This collective success, built from small incremental improvements and sustained by continuous feedback, is what transforms an ERP implementation into a strategic advantage.

Communication Strategies for ERP Success

Throughout this chapter, we have explored the human side of ERP implementations—from the need to build a resilient environment and continuously refine training programs to designing strong channels of communication and modeling change through leadership. The recurring lesson has been that technology alone cannot transform your organization. It is the people behind the technology who make the difference.

With every teacher-led workshop, every one-on-one training session and every open meeting to address system concerns, a culture is built where the ERP system is not simply tolerated but embraced as an essential tool for growth and continuous improvement.

In one example, a financial services firm discovered that initial training sessions quickly became outdated without refresher courses and peer-to-peer mentoring. As employees came to understand the nuances of the new system, they began to propose improvements that led to process enhancements and even system refinements—a clear sign that continuous learning can transform early hesitancy into proactive engagement.

This ongoing education builds not just competence but confidence, turning potential frustration into enthusiasm for new opportunities.

Communication Reduces Anxiety

This chapter also highlights the importance of clear and open communication. An ERP implementation inevitably disrupts existing workflows and, with that, may foster anxiety and uncertainty among employees. How you communicate these changes, the reasons behind them and the anticipated benefits can make all the difference.

In one project I consulted on, a large logistics company set up daily group meetings during the first month following the ERP rollout. Instead of simply delivering top-down directives, these meetings allowed every employee—from warehouse workers to senior executives—to voice concerns, share insights and collectively brainstorm solutions.

This strategy built mutual trust and created a culture in which constructive feedback became the norm. When employees feel heard and valued, they are far more likely to support an initiative and help identify opportunities for incremental improvements.

Effective Communication is a Dialogue

In parallel with training, effective communication plays a pivotal role in managing the human factor during an ERP rollout. Communication in this context is not simply about disseminating information. It is about creating a dialogue that encourages employees to share their experiences and needs.

Moreover, communication should not be a one-way street. Establish ways for employees to voice concerns, give feedback and suggest improvements.

An effective strategy employed by a mid-sized services company was to develop an online portal specifically for ERP user feedback. Employees could report issues, rate their severity and even propose solutions.

The project team reviewed these submissions weekly, providing timely responses and implementing actionable suggestions where feasible. This feedback loop streamlined system refinements and created a collaborative atmosphere where employees felt actively involved in shaping the future of their work processes.

When users see tangible outcomes from their input, their commitment to and trust in the new system naturally deepens.

I remember a large manufacturing enterprise that set up daily group meetings immediately following their ERP launch. These meetings allowed staff from every level—from front-line workers to executives—to discuss challenges openly. When issues were raised, they were addressed in a transparent manner, with leadership explaining what was being done and why it was necessary.

Over time, these open channels of communication helped build trust and dispel misconceptions about the new system. People came to understand that the

ERP implementation was not a top-down mandate to disrupt their routines, but a collaborative effort designed to make their work easier and more efficient. By maintaining a steady stream of honest dialogue, the organization reduced anxiety and turned potential resistance into a proactive search for solutions.

Clear and Open Communication

Clear and open communication is a critical pathway to success. When employees were informed about every step—from the reasons behind the changes to the practical benefits they could expect—the uncertainty and anxiety that often accompany major systems rollouts were significantly reduced.

Consider the case of a global logistics company that, facing early missteps in the system, set up daily group meetings and an online feedback portal. By sharing progress reports, addressing concerns in real time and involving employees in troubleshooting, the organization created a shared sense of purpose.

This transparency resolved issues more quickly and helped everyone see that the ERP system was not a top-down imposition, but a collaborative tool designed to ease everyday challenges and drive collective performance.

Managing Misinformation

Clear communication helps you be certain that the rationale for change, progress updates and lessons learned are shared openly throughout the organization. When employees are left in the dark, misinformation and fear can spread rapidly.

In one case involving a global logistics company, initial ERP rollout phases were marked by sporadic communication that left many staff uncertain about how the new system would impact their roles. Complaints began surfacing, and minor issues were magnified by rumor.

Recognizing the potential for long-term cultural damage, senior leadership implemented a comprehensive communication plan that stressed transparency. Regular group meetings were scheduled, the project champions hosted open Q&A sessions and updates were disseminated via an internal newsletter dedicated solely to the ERP project.

These efforts created a sense of unity and shared purpose, refocusing the conversation on how the ERP system would eventually simplify operations rather than complicate them. The logistics company's experience teaches us that clear, consistent and proactive communication is not optional. It is essential for maintaining morale and encouraging collaborative problem-solving as the project evolves.

Establishing dedicated communication channels is especially important during critical phases such as the ERP launch. For many organizations, the initial period after implementation is fraught with uncertainty as users adjust to the new interface and processes. To manage this transition effectively, companies have set up daily troubleshooting sessions and "hotlines" where employees can receive immediate assistance.

A well-documented case from a healthcare provider shows that when frontline staff were provided with a direct channel for support, the number of technical issues reported decreased over time, and overall system performance improved.

This approach resolved immediate technical troubles and reinforced the message that support is accessible and that management is committed to a smooth transition. In turn, employees feel more reassured, resulting in faster adaptation and a reduction in the overall learning curve.

Transformation and Human Adaptation

At the heart of this human-focused approach is the realization that technological transformation and human adaptation are two sides of the same coin. The ERP system, with all its capabilities, is only as effective as the people who operate it and continuously improve it.

Every time a training session is updated, every group meeting that creates insightful feedback and every leader who is visible and approachable helps reinforce the system's efficiency, and the people begin to trust and use it. This implementation does not end at launch. Instead, the launch marks the beginning of an ongoing process of optimization, where lessons from the field lead directly to system improvements and operational refinements.

In reviewing the central themes of building a resilient change culture, continuous learning and effective communication, several key points deserve emphasis.

First, ongoing training is essential. Whether through formal workshops, interactive e-learning modules or hands-on mentoring, continuous learning meets the evolving needs of the workforce and everyone is equipped to make the most of the ERP system's capabilities.

Second, open communication channels—group meetings, feedback portals, and daily cross-departmental briefings—build trust and help prevent misinformation or needless anxiety, turning potential friction into a smooth dialogue of improvements.

Third, leadership must set an example by being present, engaged and willing to adjust strategies based on real-time employee feedback. This commitment at the leadership level signals that the ERP implementation is not solely about technological change, but about a comprehensive transformation that embraces, respects and relies on the human element.

The human factor is the engine that drives the successful adoption of an ERP system. The strategies and practices highlighted here—continuous training, open communication and visible leadership—are not isolated tactics but part of a broader cultural shift. When every employee feels valued, informed and supported, the ERP system transitions into the central hub that drives operational efficiency and supports strategic decision-making.

Looking forward, the human-centric strategies will serve as an essential foundation for the next phase of your ERP implementation.

CHAPTER 7

MEASURING AND SUSTAINING ERP SUCCESS

Chapter 7: Measuring and Sustaining Success

After all the hard work in planning, executing and guiding your ERP system through the many challenges of integration and change, you now need to ask a fundamental question: Was it worth it?

In this chapter, we focus on measuring, sustaining and leveraging the success of your ERP implementation. It is one thing to put a system in place. It is another to ensure that it continually drives improvements, enhances decision-making and forms a competitive advantage for your organization.

The goal is to equip you with the tools and insights necessary to evaluate your system's performance after it goes live, adjust strategies as needed and maintain momentum through regular maintenance and iterative upgrades.

I remember working with a mid-sized manufacturing firm shortly after they had implemented an integrated ERP platform. When the system first went live, the leadership team was eager to see visible improvements. They established key performance indicators, ranging from production cycle times to inventory turnover metrics, and planned regular reviews. Within a few months, however, it became clear that not every metric was on target.

The quality of real-time data was lower than expected in a few areas, and the decision-making process was slower than hoped for—signs that the ERP system required further adjustments. Instead of seeing these issues as failures, the firm viewed them as opportunities to refine workflows and improve training programs.

This experience showed that post-implementation evaluation is not a final report card but rather an ongoing process of learning and fine-tuning. Through continuous monitoring and measurement, organizations turn an otherwise static system into a living asset that evolves alongside the business.

Post-Implementation Evaluation

Begin by examining how to conduct a thorough post-implementation review, establishing KPIs that capture the quantitative and qualitative impacts of your new system. Measuring success is not a one-time checklist but a continuous process of evaluation, reflection and refinement. This is where you turn your strategic vision into tangible benchmarks, so that the ERP system remains a dynamic asset that evolves with your business needs.

What is Successful ERP Implementation?

A successful ERP implementation is not just about deploying software—it's a transformational journey that streamlines processes, enhances collaboration and drives business growth in a competitive global landscape.

The first task in measuring success is to define what "success" really means for your organization. Before your ERP even goes live, you should have already articulated clear objectives—whether it is reducing cycle times, improving data accuracy, streamlining process flows or enhancing customer satisfaction.

For example, I recall working with a mid-sized manufacturing company that set specific targets for inventory turnover and production uptime. Initially, the leadership team expected immediate gains from the new system. However, during the first few months, they noted mixed results. There were areas that showed promising improvement, while others lagged behind.

Rather than viewing these outcomes as setbacks, they used them as a foundation for a deeper analysis to review which metrics were progressing in line with their goals and which required further attention. This proactive review helped them fine-tune the system and their workflows, transforming early discrepancies into opportunities for remedial action.

Success is marked by flexibility, agility and an unyielding commitment to process improvement—even when the results may seem modest in the short term. The overarching goal is to foster a mindset where every metric is seen as a guidepost for further improvement, every piece of feedback as an opportunity for learning and every external benchmark as a motivator for innovation.

Establish Benchmarks from the Beginning

One of the core principles in sustaining ERP success is to treat performance measurement as an iterative and holistic process rather than a fixed, one-off review. This means that from the very beginning, you should have laid out a set of clear benchmarks that reflect the operational and cultural dimensions of your organization.

For instance, an organization might start by tracking quantitative metrics such as production cycle times, inventory turnover and financial consolidation errors, while also monitoring qualitative aspects like employee satisfaction and the speed of decision-making processes.

In one illustrative case, a mid-sized manufacturing firm established baseline measurements for its operational efficiency and staff engagement right out of the gate. Over the subsequent months, monthly review meetings revealed that while their production cycle was improving, a few departments were still grappling with adapting to new data entry protocols.

Recognizing this, management refined their training approach to include more hands-on sessions and follow-up feedback, demonstrating that the process of evaluation was not only for accountability but a catalyst for continuous improvement.

Tracking the right Key Performance Indicators (KPIs) is essential to ensure your ERP implementation is delivering value. Here are the most critical KPIs to measure success in a global manufacturing environment.

Operational Efficiency KPIs

- **Production Cycle Time:** Measures the time taken from raw materials to finished goods.
- **Inventory Turnover:** Evaluates how efficiently inventory is being used.
- **Order Fulfillment Rate:** Tracks the percentage of orders delivered on time and in full.
- **Manufacturing Cost per Unit:** Monitors the cost efficiency and profitability of production.

Financial Performance KPIs

- **Return on Investment (ROI):** Assesses the financial benefits of ERP implementation.
- **Cost Savings:** Measures reductions in operational costs, waste and inefficiencies.
- **Revenue Growth:** Evaluates whether ERP contributes to business expansion.

- **Cash Flow Visibility:** Ensures accurate forecasting and financial planning.

User Adoption and System Performance KPIs

- **User Adoption Rate:** Indicates how well employees use the ERP system.
- **System Downtime:** Tracks the reliability and uptime of ERP infrastructure.
- **Process Automation Rate:** Measures reduction in manual tasks and efficiency gains.
- **Employee Productivity:** Evaluates improvements in workforce efficiency.

Supply Chain and Customer Satisfaction KPIs

- **Supplier Lead Time:** Monitors how quickly suppliers deliver materials.
- **Customer Satisfaction Score (CSAT):** Measures customer experience, post-implementation.
- **Order Accuracy:** Ensures fewer errors in manufacturing and order processing.
- **Demand Forecast Accuracy:** Improves production planning and resource allocation.

Compliance and Risk Management KPIs

- **Regulatory Compliance Rate:** Tracks adherence to industry and legal standards.
- **Data Accuracy:** Ensures the integrity and accuracy of business-critical data.
- **Security Incidents:** Measures occurrences of data breaches or unauthorized access.

A successful ERP system should enhance visibility, efficiency and profitability while ensuring seamless collaboration between departments.

Use External Benchmarking

Beyond internal processes, external benchmarking also plays an essential role in measuring ERP success. Comparing your performance with industry standards or competitors' metrics helps to contextualize your achievements and pinpoint areas for further improvement.

I recall a case involving a logistics company that, after rolling out its ERP solution, discovered through external benchmarking that its order fulfillment times lagged behind industry averages. This realization prompted the company to re-examine its internal processes and its interactions with its supply chain partners.

As a result, they established new protocols for data sharing and vendor collaboration that led to a marked improvement in performance. In this way, external benchmarks serve as a reality check, ensuring that the system's gains are competitive and forward-thinking. Evaluating your system's performance against industry standards is an objective measure of progress and creates a fresh perspective on areas for improvement.

For example, a company in the consumer goods sector compared its ERP-derived metrics—such as order cycle times and vendor performance—with those of leading peers in the industry. This external analysis revealed that, while they had made significant progress internally, there was still a gap in the timeliness of supply chain data.

With this knowledge, the company initiated a dialogue with its vendors and re-engineered certain aspects of its data exchange process, leading to better alignment with industry best practices.

Such external validation boosts confidence and ensures that your ERP system remains relevant and competitive in a broad market context.

Post-Implementation Review

You need to know how to conduct a thorough post-implementation review. Measuring success begins with clear, relevant metrics that capture the quantitative and qualitative impacts of the ERP system. These metrics must consider immediate improvements—cost reductions, improved cycle times and improved data accuracy—and also capture subtle shifts in organizational culture and employee satisfaction.

In one case, a manufacturing company set out to monitor customer service metrics alongside operational KPIs. When they found that employee feedback indicated a reduction in daily stress and a greater sense of ownership in their roles, it validated that success was measured in more than just dollars saved.

This balanced perspective allows you to understand that long-term success includes the whole of the organization—systems, people and processes—working in harmony.

Once you have established your metrics, you need to create a systematic process for regular evaluation. A rigid "once and done" approach seldom captures the ongoing dynamics of an ERP system. Instead, scheduling periodic reviews can help identify early warning signs and create guidance for tweaks before minor issues compound into significant problems.

I recall a technology company that, after implementing its ERP, initiated monthly review meetings with representatives from each department. These meetings served as a review, where operational data was compared with expectations, and lessons were gathered from any discrepancies.

The feedback collected in these sessions was then funneled into strategic adjustments that drove continuous improvements across the enterprise. This iterative process ensures that the ERP system remains aligned with changing business needs and that any necessary adjustments are made promptly.

Schedule Regular Evaluations

Once KPIs are identified, establishing a systematic review process is critical. Many organizations have learned the hard way that a "set-and-forget" model leads to stagnation and missed opportunities. Scheduling regular evaluations—be they monthly, quarterly, or at other strategic intervals—ensures that emerging issues are detected early.

In one memorable project, a global electronics manufacturer set up a series of structured review meetings, during which representatives from finance, operations and IT would come together to compare actual performance against target metrics. These meetings were not just status updates; they served as platforms for deep diving into the root causes of any performance gaps.

For example, if production cycle times were not improving as expected, the team would examine whether it was a systemic issue with data entry, a misalignment between scheduling software and the shop floor or perhaps a training gap among frontline staff. The willingness to confront these issues head-on and adapt the system accordingly was a key factor in sustaining long-term success.

Does ERP Induce Innovation?

Another perspective to consider is how the ERP system drives innovation within your organization. For many, ERP implementation is viewed primarily as a tool for operational efficiency; however, the strategic value of an ERP system is also found in its ability to open new avenues for business development.

For instance, one of my clients, a fast-growing manufacturing company, used its integrated ERP data to identify market trends and optimize inventory management. By correlating production data with sales figures in near real time, they were able to adjust their forecasts more accurately and negotiate better terms with suppliers.

This proactive use of data enhanced operational agility and positioned the company to seize new market opportunities with confidence. In this sense, the ERP system becomes more than a tool for sustaining operations, it becomes a strategic asset that fuels innovation and competitive advantage.

Sustaining ERP Success

Sustaining the success of an ERP system is an ongoing effort that requires constant vigilance, flexibility and the willingness to adapt as your organization and market conditions evolve. It is about ensuring that the system remains an engine for continuous operational improvement and strategic insight. Your measurement practices, continuous feedback loops and data-driven decision-making processes all converge to create an environment where the ERP is not a static platform but a dynamic, integral part of your organization.

In this section, we go deeper into the systematic approaches for actively monitoring results, refining processes and learn how to be certain that the ERP system remains a dynamic engine driving long-term strategic growth.

Building upon the initial establishment of KPIs and feedback loops discussed earlier, this section focuses on detailed strategies for continuous monitoring, organizational realignment and leveraging data-driven insights to maintain a competitive advantage.

Successfully implementing an ERP system is a significant achievement, but the true challenge lies in sustaining that success after deployment. ERP is not a one-time project. It's an evolving asset that requires continuous optimization, user engagement and strategic adaptation. Business leaders must cultivate a dynamic approach to ensure their ERP solution remains a driver of innovation, efficiency and growth.

ERP is far more than just a system. It's a dynamic asset that evolves alongside a business, driving efficiency, scalability and innovation. For business leaders spearheading ERP implementation, understanding its adaptability is crucial. A well-integrated ERP solution streamlines operations by unifying core business functions, providing real-time insights and enhancing decision-making. But its true power lies in its ability to grow with an organization, accommodating new processes, market shifts and technological advancements.

Leaders must champion an agile mindset, ensuring ERP remains a catalyst for progress rather than a static tool. By fostering user engagement, continuous optimization and integration with emerging technologies, businesses can transform ERP into a strategic advantage that empowers long-term success.

Process Refinements

At its core, a successful ERP system is one that meets initial implementation goals and adapts to the evolving needs of the business. It becomes a platform for ongoing innovation and a source of actionable intelligence that supports strategic decision-making.

The certainty and accuracy of integrated data, coupled with constant engagement from all stakeholders, create an environment where the ERP solution continuously propels the organization forward.

The value of measuring and sustaining success lies in transforming the ERP from a technical installation into a strategic asset that drives competitive advantage and fuels long-term growth.

A particularly illuminating example comes from a global manufacturer that had successfully implemented its ERP system and invested heavily in post-implementation reviews. Within the first year, they improved operational metrics such as production throughput and inventory turnover and identified new opportunities for cost reduction that had not been foreseen during planning.

By closely monitoring these outcomes and adjusting its internal processes accordingly, the company achieved an impressive return on investment. This case reinforces the idea that the journey does not end at implementation. In fact, it was only the beginning of a long-term process of continuous improvement and strategic refinement.

Ensuring the long-term success of an ERP system requires more than a one-time launch. It demands an ongoing commitment to measurement, adjustment and strategic alignment. With the establishment of clear metrics to evaluate performance, regular feedback loops for lessons learned, a culture of continuous training and a proactive approach to system maintenance and upgrades, the ERP platform can evolve into a perpetual resource for growth and innovation.

As you reflect on these practices, consider the integration of robust post-implementation reviews, ongoing training and strategic updates into your overall ERP management plan. This holistic approach ensures that your system serves immediate needs and adapts fluidly as your organization grows and markets change.

With these thoughts in mind, you are well-equipped to progress from measuring the impact of your ERP implementation to fully leveraging its capabilities to drive competitive advantage and achieve long-term operational excellence.

Regularly Revisiting ERP Strategy

Beyond internal adjustments, sustaining success involves regularly revisiting your ERP strategy to see if it aligns with your long-term business goals. Market trends, competitive dynamics and even internal growth can shift the relevance of specific ERP features over time.

Organizations that schedule regular strategy reviews are in a better position to reallocate resources, retool processes or even implement system upgrades that align with new opportunities.

For instance, a company in the logistics sector used quarterly strategy reviews to examine operational data and to forecast potential market shifts. When they noticed that their system's reporting capabilities were lagging behind industry demands, they proactively worked with their vendor to enhance those modules. This agile approach ensured that the ERP system became a dynamic component of their overall strategy rather than a static fixture.

Continuous Learning

Another critical insight revolves around the importance of integrating continuous learning into your ERP strategy. Successful organizations understand that no system, no matter how advanced, stands still. The environment in which your ERP system operates is constantly changing. New market conditions emerge; business processes evolve and the competitive landscape shifts. Consequently, ongoing user training is essential. Rather than viewing training as a one-time event that concludes at launch, the successful companies built comprehensive, ongoing education programs that catered to different user roles and kept pace with system updates.

For example, the financial firm I examined implemented an online training portal and periodic refresher sessions that improved product usage and gave employees confidence to suggest further enhancements.

In the end, continuous training supports a culture where every team member is equipped to maximize the ERP system's capabilities, ensuring that the benefits of your investment are sustained through a skilled and adaptable workforce.

Open Communication

Equally important is the role of open communication, which emerged as a central pillar throughout our discussions. Detailed, transparent communication channels—ranging from regular group meetings to structured cross-departmental review sessions—are critical in maintaining alignment across the organization.

When employees understand why changes are being made and witness that their feedback leads to concrete improvements, resistance diminishes and collaboration flourishes.

In one case, a global logistics company used a combination of daily operational briefs and an internal online portal for feedback, which allowed them to quickly identify issues such as data entry delays or discrepancies in inventory reporting. This open dialogue helped the organization resolve these issues in real time while also building a shared sense of ownership over the system's ongoing success.

The success of these initiatives reinforces the notion that communication should be two-way. Listening, as much as reporting, is key to sustaining the system's performance.

Risk Management

Risk management has also proven to be a critical component in sustaining ERP success. By establishing a robust risk register early on and updating it regularly, organizations can anticipate potential setbacks before they evolve into larger problems. Even carefully laid plans run into unexpected challenges—technical setbacks, external vendor delays and shifting market conditions.

By maintaining a proactive risk management strategy, organizations can reallocate resources, adjust timelines and reconfigure processes to mitigate the impact of these uncertainties.

For example, one logistics company I consulted for had a comprehensive risk register that highlighted potential disruptions during peak seasons. When warnings arose, they adjusted staffing and recalibrated production schedules, accordingly, ensuring that operations continued with minimal disruption. This continuous risk assessment process is not simply about avoiding problems. It is about creating an agile culture that sees every challenge as an opportunity for improvement.

Security

Security and compliance also come into play when we consider advanced usage. As your ERP system reaches deeper into business processes and interacts with external partners or advanced analytics modules, the perimeter of risk grows. Regulatory requirements, data privacy concerns and the threat of cyberattacks become increasingly relevant.

Consequently, embedding security protocols in your advanced ERP usage is a necessity, not a luxury. Monitoring user access, encrypting sensitive data, enforcing vendor compliance and regularly conducting security audits must remain core practices.

A healthcare organization that added telemedicine integration to its ERP discovered that each new interface introduced potential vulnerabilities, requiring strong authentication, encryption and audit logs. They systematically tested these new modules with mock data breaches and continuous threat assessments, so that the system remained secure while maintaining the operational benefits. The advanced capabilities provided new patient care options, but that success hinged on responsibly managing the expanded risk landscape.

Brand Identity

A decisive point to consider in this advanced stage is how your ERP approach can shape your organization's brand identity. When clients, partners and market observers perceive that your systems, processes and data-driven culture consistently deliver reliability, responsiveness and innovation, your brand equity naturally rises. ERP systems that show strong analytics and streamlined collaboration can be turned into marketing points.

A distribution company, for example, highlighted its real-time inventory tracking and rapid response to shipping disruptions in sales presentations, effectively demonstrating operational excellence that influenced prospective customers' trust. The brand story became one of reliability and technology-forward thinking, a narrative strongly backed by the ERP system's operational maturity.

Pulling these threads together, leveraging an ERP system for strategic advantage necessitates more than just stable operations and neat dashboards. It calls for a determined shift toward proactive decision-making, advanced analytics adoption, deeper internal and external collaboration, reevaluation of leadership roles, strong security protocols and an overarching goal of shaping brand identity through operational excellence.

The "how" behind these transformations rests on everything you have learned thus far, which includes, well-maintained data governance, continuous training within a supportive culture, consistent feedback loops and a strong leadership commitment.

As you look ahead to the outcome of your ERP implementation, ask yourself how each advanced initiative—be it predictive analytics, integrated business intelligence, or vendor collaboration—ties back to your organizational strategy. Because, in the end, technological choices are only as effective as the strategic governance under which they operate.

The best examples of ERP success—from the manufacturer that seized an early-market opportunity in smart devices, to the manufacturer that harnessed real-time data for agile decision-making—happened because leaders connected system-enhanced insights directly to business objectives. This alignment transforms an ERP system from a powerful operational tool into a strategic asset capable of reshaping entire industries if used to its fullest extent.

Throughout these reflective narratives and guiding principles, one message remains constant, which is, success depends on continuously evolving not just the system itself, but the mindset of every stakeholder. The moment your organization sees the ERP platform as a living, central nervous system for daily operations and forward-looking strategy, you free yourself from the constraining notion that technology projects have discrete start-and-end points.

Instead, your view broadens to see endless possibilities—expanding market reach, unveiling new revenue streams, strengthening brand value and even pioneering novel business models. All of these revolve around the same core of well-integrated data, strong analytics and a corporate culture that welcomes change.

The lessons gleaned from these examples will bring to life the advanced scenarios where your ERP system can push boundaries and deliver results far beyond the usual operational upgrades.

ERP technology is foundational, but the real magic emerges from how well its data is harnessed, how openly people collaborate around that data, how significantly leadership invests in advanced functionalities and how effectively you integrate new insights into the organization's strategic direction.

Use this moment to take stock of your ERP environment's readiness for these advancements. Does your data governance regime support advanced analytics, insights and external integrations? Do your staff have the skills and the frameworks to act on data insights swiftly? Are business strategies and technology roadmaps in sync for the next wave of deployments?

Each positive answer bolsters your capacity to stand out in the market and sustain that advantage. Each gap highlights where further refinement might be necessary. By systematically bridging these gaps, you empower the ERP system to move from an adept operational platform into a strategic powerhouse, one that responds to the present and shapes the future.

Evolve with Business Paradigms

The most compelling insight from this chapter is that sustaining the success of an ERP system is a never-ending process. The system itself is not a final product but an evolving element of your organization's infrastructure. Continuous improvement is achieved through regular reviews, constant feedback, ongoing training and proactive risk adjustments. Every time you refine a process or update training modules, you ensure that the system evolves along with your business.

A global manufacturer I collaborated with set up quarterly reviews that led to refinements in their data integration protocols and user interfaces, which then translated into measurable improvements in production efficiency. These iterative cycles are what transform the ERP system into a viable asset that continuously contributes to your organization's growth and competitive advantage.

The key components for sustaining ERP success include:

- Establishing clear, comprehensive and balanced KPIs that capture quantifiable and qualitative impacts.
- Commit to continuous learning and training that adapts as the system and business evolve.
- Create robust, multi-directional communication channels that facilitate transparency and engagement across all levels.
- Implement proactive risk management practices that keep you ahead of potential disruptions.
- Practice continual evaluation and external benchmarking to ensure your ERP system remains competitive and aligned with strategic goals.

These elements work together to create an environment where the ERP system is maintained and continually optimized, evolving to support new business challenges and opportunities.

When your organization adopts a mindset of ongoing improvement, every performance review and every risk mitigation plan supports operational excellence. Even when market conditions shift or internal processes change, you will have built the resilience required to adjust quickly and effectively.

In one illustrative case, a mid-sized manufacturing firm that initially struggled with data integration issues chose to document every performance metric and update its risk register monthly. Over time, they discovered recurring patterns that led to targeted training updates and adjustments in system configuration.

This disciplined, iterative approach resolved the initial issues and paved the way for further innovation, such as dynamic scheduling and improved vendor management. The integrated feedback from users, combined with external benchmarking, ensured that the ERP system grew more robust with every iteration.

Sustaining ERP success is a multifaceted process that combines diligent measurement, continuous learning and ongoing adjustment. It requires a proactive stance on performance evaluation, a willingness to embrace feedback from every stakeholder and a flexible strategy that adapts to the evolving business landscape.

When combined, these elements transform the ERP system from a static installation into a dynamic, strategic asset that achieves immediate efficiency gains and supports long-term growth and competitive differentiation.

As you move forward, remember that the ERP system's long-term benefits lie in its adaptability and integration into your overall business strategy. By maintaining rigorous evaluation practices and fostering a culture of continuous improvement, your ERP system will remain an invaluable resource capable of supporting your organization's strategic objectives over the long haul.

The lessons learned from constant monitoring and refinement feed directly back into the business's broader strategic planning. They ensure that the system's evolution is aligned with changes in the market and internal organizational shifts. With continuous input from experienced staff, open feedback channels and solid external benchmarking, you create a comprehensive process system wherein the ERP system serves as a tool for daily operations and as a driver of innovation and strategic agility.

By now, you should appreciate that sustaining ERP success goes far beyond the technical aspects of system installation. It is an ongoing process of measurement, learning and adaptation. The insights gathered from the various strategies illustrate that an ERP system's effectiveness hinges on ongoing attention to detail, which includes a dedication to revisiting performance metrics, adapting training programs and being agile enough to respond to unforeseen challenges. The culture of continuous improvement, built on robust communication channels and supportive leadership, transforms initial successes into enduring competitive strengths.

As you prepare to transition to the next phase, which is dedicated to leveraging your ERP system as a continual strategic advantage, it is essential to view the processes of measurement and sustained improvement as ongoing commitments rather than isolated steps.

With these comprehensive insights into sustaining success, you are now well-equipped to harness the full power of your ERP system. Every process refinement, every feedback review and every strategic adjustment becomes a building block for a robust operational framework that propels your organization forward. This process of continuous improvement ensures that your system evolves organically with your organization, supporting immediate operational gains and long-term strategic objectives.

As the focus shifts to leveraging this operational excellence into a concrete competitive advantage in the following phases, remember that the continuous, iterative process of sustaining success is the key to turning your ERP investment into a lasting catalyst for improvement and innovation.

ERP is an Essential Strategic Asset

Sustaining ERP success is about forging a long-term relationship with your ERP system that is built on measurable performance, continuous learning and adaptive strategies. This ongoing commitment to excellence transforms the ERP solution from a one-time project into an enduring platform that drives your organization's competitive edge and operational efficiency.

Every process improvement, every update driven by employee feedback and every proactive risk management action contributes to a culture where the ERP system remains a critical strategic asset.

As we conclude our exploration in this chapter, let's summarize these core insights:

- **Clearly defined KPIs** that include quantitative and qualitative measures are essential for sustaining ERP success.

- **Continuous learning**, rather than one-off training sessions, enables employees to keep pace with changes in the system and contributes to a culture of ongoing improvement.

- **Open and structured communication channels**—group meetings, feedback portals and cross-functional meetings—are critical for maintaining trust and collective ownership.

- **Proactive risk management**, involving detailed risk registers and regular assessments, helps the organization mitigate challenges before they escalate.

- **External benchmarking** creates context for your internal performance and ensures that your ERP system remains competitive in a broader market.

- **Evolutionary paradigms** require viewing the ERP implementation as an evolving process, where ongoing reviews and iterative improvements transform the system into a dynamic strategic asset.

These practices collectively ensure that your ERP system continues to deliver value long after its initial deployment, transforming from a static tool into a continuously improving platform that aligns with your strategic vision. By building a culture that embraces ongoing evaluation and refinement, you enable your organization to not just adapt to changes but to lead in a competitive market landscape.

As you consider these ideas, think about how the rigorous measurement, continuous learning and adaptive strategies discussed here can be embedded into your overall business planning. The ERP system, when managed through an ethos of continuous improvement, does more than simply support daily operations. It becomes the central nervous system that informs every strategic decision, every resource allocation and every innovation initiative.

Sustaining ERP success is a perpetual process that requires dedication, agility and a collaborative spirit. The system evolves in tandem with your business, shaped by every review, every feedback session and every adjustment.

CHAPTER 8

ERP FOR COMPETITIVE ADVANTAGE

Chapter 8: ERP for Competitive Advantage

Using ERP for competitive advantage is about harnessing the system's full potential as a strategic force that can guide your organization's competitive positioning and long-term growth. In earlier chapters, we addressed each step—selecting the right solution, lining up accurate requirements, calibrating customization levels and managing the complexities of training, change management and continuous improvement.

By now, you have seen how ERP projects can challenge entrenched processes and test an organization's adaptability. You have also learned that just implementing software is not the end of the road; rather, it is a starting point for ongoing refinement, innovation and strategic use of data-driven insights.

As you stand on the threshold of this ever-evolving phase, the question changes to—How do we leverage this system to achieve a significant competitive advantage?

Whether it is pushing for operational agility, cultivating advanced analytics, or expanding into new markets, your ERP system will serve as the backbone and the catalyst for evolving strategic directions. The aim is to show how superior data visibility, improved decision-making processes and a well-integrated technology platform can open doors to growth, efficiency and customer impact that go well beyond the original scope of the implementation.

As we move forward in this chapter, our focus will be on synthesizing the core principles that have emerged from every facet of the ERP implementation process. We will discuss how to embed these learnings into your organizational strategy, ensuring that your ERP system is an operational tool and a catalyst for continuous improvement and innovation.

This synthesis is a launchpad for a transformed perspective that takes your ERP system beyond the realm of technology and positions it as a central component of your strategic future.

Our discussion will also address the importance of planning for the inevitable future upgrades and adaptations. The business environment is in constant flux—new technologies emerge, market conditions shift and competitive pressures demand agility.

Every stakeholder within the organization plays an essential role in the long-term success of the ERP system. The cumulative insights—from setting clear KPIs and engaging in continuous training to maintaining open communication and proactive risk management—form the bedrock upon which sustainable operational excellence is built.

An ERP system that evolves through a culture of continuous improvement is an asset for today, and a strategic investment for the future.

Let us now prepare for the transition, where we will focus on how to leverage your fully integrated ERP system to drive a transformative, future-proof strategy. The synthesis of planning, execution, human engagement and continuous improvement does not mark an endpoint but a beginning—a launchpad for the next stage of evolution, where innovation meets operational excellence, and every insight is transformed into strategic accomplishment.

By measuring, sustaining and continuously improving your ERP system, you are equipped to carry forward the legacy of your implementation. The foundation you have built through rigorous evaluation, persistent training and adaptive communication will allow you to harness your ERP system as a competitive advantage.

With these insights firmly in place, you stand ready to move to the next stage dedicated to leveraging your ERP implementation as a competitive engine, where each refined process, enhanced metric and strategic pivot combines to drive your organization into a future of sustained excellence. This concluding phase will explore the actionable strategies for transforming your integrated ERP system into a strong catalyst for innovation and market leadership.

Now, with the experience and knowledge accumulated from sustained success, let us use every lesson learned to refine, leverage and propel your organization forward.

Orchestration of Technology, Processes and People

At its core, ERP implementation is much more than software installation. It is a holistic transformation that touches every part of an organization—its technology, processes and, importantly, its people. Over the course of this book, we have discussed how choosing the right system, forging a balanced relationship between customization and standardization and managing the myriad challenges of implementation all set the stage for operational excellence. Now, to consolidate those insights, the goal is to help you reflect on the process, celebrate the progress made and prepare for the future. In simple terms, it's about practicing the orchestration of technology, processes and people.

As you reflect on the collective wisdom of these discussions, consider the multifaceted nature of sustainable ERP transformation. It is not enough to install a system and hope for automatic improvements. Instead, the long-term value of your ERP investment depends on embracing change as an ongoing process.

Every training session, every feedback loop and every proactive risk management activity contributes to a culture where innovation is continuous, and adaptation is the norm. The ERP system, when nurtured and evaluated regularly, becomes an integral part of your strategic planning, informing financial decisions, driving process improvements and even positioning your organization competitively in the market.

This chapter is an opportunity to tie together all these strands into a coherent narrative that reviews what has been achieved and sets the stage for the future. Recognize the importance of a disciplined approach, which includes clear requirements, thorough vendor evaluations, balanced customization choices and the willingness to face technical and human challenges head-on.

Each phase, from preparation to execution, from post-implementation measurement to ongoing refinements, forms a critical link in the chain of ERP success.

I recall the story of a manufacturing firm that initially struggled with data inconsistencies, which led to several months of troubleshooting. Instead of being discouraged, the team documented each setback, refined their data protocols and engaged in continuous training until the system was performing at a level that enabled the company to regain its competitive edge.

Shift from Reactive to Proactive Strategies

One of the central themes here is the shift from reactive to proactive business strategies. We explored scenarios where real-time dashboards and automated reports allowed organizations to spot issues quickly and resolve them before they turned into systemic bottlenecks. In this phase, the focus is not just on managing challenges but on anticipating opportunities.

A mature ERP system allows a business to draw from integrated data sources to forecast emerging market demands, identify profitable customer segments and confidently adapt its product or service lines.

For example, a global logistics company I once consulted had initially adopted ERP software to unify its shipping, warehousing and billing processes. After successfully stabilizing operations, the company used the system's extensive data streams to analyze underutilized routes and explore new partnership opportunities, transforming the original goal of streamlining processes into a forward-looking strategy that expanded their market reach.

Success metrics include the strategic adoption of advanced analytics and intelligence capabilities. Throughout this book, I have repeatedly stressed the importance of reliable data. Yet data in isolation only goes so far. The real power surfaces when that data is continuously translated into actionable insights.

Advanced Analytical Capabilities

One way of leveraging your ERP for maximum advantage is adopting and refining advanced analytical capabilities. While real-time reporting and dashboards are effective, many organizations find that they gain exponential benefits by integrating advanced data tools—such as machine learning, predictive analytics and AI-driven insights—directly with their ERP environment.

Many ERP platforms today offer or integrate with analytics modules to incorporate predictive modeling, machine learning and sophisticated visualization tools. This integration allows for collaborative intelligence. Marketing teams can unearth new trends in customer behavior. Operations can predict inventory demands more accurately and finance can model best-case and worst-case scenarios with relative ease. The results can dramatically enhance agility.

One organization we observed used advanced analytics within its ERP environment to pinpoint inefficiencies in product distribution. By correlating shipping data with regional sales forecasts, they optimized shipping schedules and cut warehousing costs by nearly 20%. This deeper analysis would have been unattainable without a strong, integrated ERP platform functioning as the central repository and processing hub.

Then, there is the principle of aligning the ERP system tightly with broader strategic goals, short- and long-term. Previously, we examined the necessity of tying performance metrics, staff training and system features to the organization's daily operations.

In a fast-evolving market, the ability to generate forecasts, discover hidden trends or automate repetitive analytical tasks can mark the difference between simply keeping pace and leading the charge. For instance, a logistics company that integrates predictive analytics into its ERP can forecast potential delivery surges around peak shopping seasons, head off capacity shortages and coordinate staff scheduling more intelligently.

Data that was once locked in spreadsheets or siloed in various modules can be aggregated, analyzed and converted into actionable directives. This approach streamlines operations and creates a data culture where employees from various disciplines learn to rely on strong, evidence-based reasoning.

Enabler of Strategic Initiatives

In this chapter, we go one step beyond. Your ERP system, at its peak usefulness, should no longer be viewed as a standalone technology, but as an enabler that shapes and executes strategic initiatives.

Consider a manufacturing company that once focused solely on improving inventory management to reduce costs. Having refined the system over time, leadership began using real-time sales and supply data to guide decisions about store locations, product diversification and even potential acquisitions of smaller manufacturers.

Their ERP software informed daily processes and actively influenced major strategic choices that reshaped the company's growth trajectory. This synergy between technology and organizational vision exemplifies what it takes to gain a lasting competitive edge.

Remember, an ERP system that evolves with your business can also encourage new forms of collaboration, internally and externally. Within the organization, cross-functional teams, armed with integrated data, can work more transparently and effectively, generating creative solutions to complex problems.

Across the supply chain, vendors and partners can also connect with your system under carefully managed permissions, allowing real-time data exchange that reduces delays and eliminates redundant steps.

An international electronics manufacturer I encountered used open ERP interfaces to integrate supplier status updates, enabling them to quickly spot variations in supply and pivot production schedules. This close partnership between the manufacturer and the supplier elevated trust, cut lead times and reinforced the idea that the ERP system's benefits can spread beyond the corporate walls.

A similar integration is possible when supporting the needs of your customers. Supporting a just-in-time inventory fulfilment strategy is an example of the need to coordinate ordering and deliveries for your customers. Extending ERP to work with data from your customers allows the just-in-time inventory replenishment process to be automated.

In a swiftly shifting market environment, these collaborative networks become a key differentiator, creating the kind of synergy that can be difficult for competitors to replicate.

At this stage, leadership and governance remain essential. Although much of the earlier work may now feel routine, such as ongoing training, performance reviews and risk management, leaders must keep the ERP system visible in strategic conversations.

By showcasing the system's ongoing contributions and highlighting new opportunities for leveraging data, top management can sustain interest, secure needed investments and keep your system from stagnating.

One leadership team I worked with made it a habit to feature "ERP-driven successes" in weekly bulletins and monthly board meetings. This consistent emphasis reminded every stakeholder of the system's strategic relevance, thus avoiding complacency and underuse.

A carefully governed ERP system, with an active and motivated leadership team behind it, naturally matures into a competitive instrument that helps shape and guide the future rather than only document the present.

Another dimension to consider here is innovation and the role your ERP system plays in driving new business models, products or services. The continuous streams of real-time data that well-implemented ERP solutions provide can serve as the foundation for product ideation sessions, R&D planning and market trials. In many technology-driven markets, being able to pivot swiftly is often the key to survival.

A mid-sized software company once used the data from its ERP system to identify usage patterns in product distribution, quickly adapting its licensing model to cater to a rising segment of small-business users.

Without the thoroughness and accuracy of real-time data, made accessible by a well-configured ERP, the company might have missed this emergent market or caught on too late to reap the benefits of first-mover advantage. Their ability to innovate was directly tied to the system's capacity for providing timely, relevant insights.

Amid these compelling narratives of strategic leverage, it is vital not to overlook the necessity of continuous staff development. Even as the system matures, new features, updates and expansions will roll out, whether from the vendor or internal development teams. This means training cannot stagnate.

Bridging the Competency Gaps

The most effective organizations persistently re-examine user competencies, bridging any gaps either through formal learning modules or structured peer mentorship.

I recall a healthcare provider who integrated a new analytics module for patient data after the initial ERP launch. Initially, only a few super-users were trained to manage the advanced reports.

Over time, the leadership recognized that the entire organization could benefit from these powerful analytics. They then rolled out additional user training across various departments, enabling a broader user base to harness the data for improved patient care and streamlined administrative tasks.

The result was a more inclusive, innovation-driven environment, with the ERP at its core—an environment where every employee, from administrative staff to medical practitioners, found real value in the system's expanding capabilities.

As you explore ways to transform your ERP system into a linchpin of strategic advantage, you will likely revisit many of the frameworks and lessons covered in previous chapters. The principle of balanced customization continues to hold merit.

Going overboard on tailor-made features might hamper future upgrades or expansions. Meanwhile, the discipline of risk management helps identify potential vulnerabilities early as your system branches out to new functions and partner integrations.

Moreover, the feedback loop approach remains essential. You must maintain and even amplify the channels through which employees share insights and improvements to keep your ERP system agile. Each step, from integrating external partnerships to implementing advanced analytics, must be tested against real-world feedback to confirm that it truly adds value and aligns with strategic objectives.

The process of leveraging ERP for genuine, sustained competitive advantage requires a dual mindset. You must remain grounded in daily operational realities while constantly scanning for new opportunities to innovate and improve.

The system's data can alert you to inefficiencies, point to fresh strategic directions and even highlight changing consumer or vendor behavior, which can inform your next move in the market. An ERP solution, managed and well-nurtured, becomes an integral part of how your organization thinks, plans and grows.

The pivotal themes of this stage are using the ERP to spur proactive strategy, adopting advanced analytics and intelligence capabilities, aligning the system with overall business goals, fostering deeper collaboration inside and outside the enterprise and ensuring leadership continues to champion the system's evolution.

Try to see beyond the operational improvements you achieve to envision your ERP system as a gateway to ongoing strategic transformation. By embedding advanced analytics, forging stronger alignments with your broader goals and championing an organization-wide culture of integration and innovation, you raise the ERP system from a supportive role to a central function in your success story.

Unlocking Growth Trajectories

Reflect on how your ERP system can shift from simply generating operational efficiencies to unlocking new growth trajectories.

Drawing on the lessons gleaned from each phase of this process—from the painstaking vendor selection to the delicate balancing of custom and standard features, from the nuanced guidance of human change management to persistently updated metrics—your organization stands on a firm footing, ready to explore the advanced possibilities that a maturing ERP platform can offer.

With each improvement and each new integration, you position yourself to make bolder strategic decisions, respond faster to market changes and spearhead innovations that can differentiate you from the competition.

The aim of these concluding discussions is to empower you to look beyond routine maintenance and daily usage to see ERP as a conduit for big ideas and for truly transformative thinking that can propel your enterprise into new arenas or help redefine existing ones.

Keep in mind, as always, that at the center of this evolution lies your organization's culture that encourages continuous learning, invites cross-departmental collaboration and promotes open communication. No matter how sophisticated ERP technology becomes, continued success lies with the leaders and employees who shape its use. When that culture is firmly in place, the path to leveraging ERP for strategic gain becomes clearer and far more achievable.

By focusing on proactive strategy, advanced analytics, strong alignment with business goals, enhanced collaboration and unwavering leadership commitment, you build a framework for harnessing your ERP system as a vital force in your organization's forward momentum.

Launchpad for Competitive Strategy

An ERP system's full value is realized when it does not simply automate existing processes but enables new ways of working that drive competitive advantage.

For instance, one of my clients, a fast-growing manufacturing company, found that the true benefit of their ERP system was not just streamlining production schedules and its ability to foster better cross-departmental collaboration.

With access to integrated data and real-time dashboards, management was able to identify market opportunities faster than competitors. Their system created insights that led to process innovations, such as dynamic inventory management and improved vendor negotiations, which then delivered tangible financial benefits.

This example demonstrates that the ERP system should serve as a launchpad for competitive initiatives, enabling strategic decisions that keep your organization ahead in ever-changing markets.

Align ERP with Strategic Business Plans

In addition, sustaining long-term success involves aligning the ERP system's output with your strategic business plans. A well-integrated system should enable leadership with actionable insights that contribute directly to shaping the organization's future.

In one illustrative case, a fast-growing manufacturing firm used its ERP data to reassess its production forecasts, identify slow-moving inventory and adjust its sourcing strategies accordingly.

The subsequent improvements in resource allocation and production planning drove operational efficiency and resulted in a measurable boost in profitability, showcasing how a well-maintained ERP system can directly contribute to strategic success.

Another example is a scenario involving a consumer goods company that relied heavily on its ERP reports for strategic decision-making. Initially, the executive team was dissatisfied with the timeliness of the insights.

Upon digging deeper, they realized that the system's reporting functionalities were not fully optimized for real-time data analysis. By collaborating with their vendor to upgrade the analytical tools, they transformed their decision-making process.

The improvements in data quality and timeliness created a clear competitive edge, underscoring the idea that an ERP system's long-term value lies in its capacity to continuously inform and support strategic decisions.

Show Key Metrics on Performance Dashboards

In practice, sustaining success requires a robust framework for data collection and analysis. This means embedding performance dashboards and automated reporting tools within the ERP system itself. When key metrics are readily accessible in real time, decision-makers at every level can identify trends, spot potential issues early and act before minor disruptions grow into significant problems.

One global electronics manufacturer I worked with installed interactive dashboards that aggregated data from multiple departments into a single view. These dashboards allowed the leadership to see, at a glance, that while throughput had increased overall, there were recurring bottlenecks in the quality control process.

Armed with this insight, the company initiated targeted improvements in that area, adjusting workflow timings and allocating additional resources. The result was a smoother production process with a renewed focus on quality that enhanced the firm's reputation in the market.

Reward System Performance

For many organizations, sustaining momentum involves ensuring that everyone—from frontline staff to senior management—remains engaged with the ERP system long after implementation. One effective method to encourage continuous engagement is to link system performance directly to reward systems and performance reviews.

In one organization, employees who contributed ideas that led to measurable process improvements were recognized and rewarded, creating a culture where continuous improvement became a shared priority. This kind of recognition boosts morale and reinforces that every individual can contribute meaningfully to the organization's ongoing success.

Maintain a Performance Register

Maintaining a dynamic performance register that is revisited at regular intervals can help pinpoint areas where the ERP system may be underperforming or where external factors have altered the competitive landscape.

For instance, a logistics company that I advised set up a detailed performance matrix during implementation, which was then constantly updated with input from field teams. When an emerging trend indicated a potential delay in order processing during peak seasons, the management preemptively adjusted staffing levels and recalibrated their delivery models.

Such anticipatory actions mitigated potential losses and reinforced the ERP system's ability to serve as a guiding tool in a swiftly changing environment.

Collaborative Problem Solving

A critical element to sustaining success through measurement is the establishment of regular, structured review sessions that bring together representatives from across the organization. These sessions should be more than just status updates. They are forums for collaborative problem solving, where insights from quantitative data are discussed in light of operational experiences.

For example, a large manufacturing company implemented monthly cross-functional reviews, during which department heads, IT specialists and frontline employees analyzed performance metrics together. In one such meeting, it was revealed that discrepancies in inventory data were leading to recurring replenishment delays—a seemingly minor issue with a significant cumulative impact.

The team used this opportunity to reexamine the data flow from warehouse sensors to the central ERP system and fine-tune the integration points to ensure better accuracy. By opening these channels, the organization fostered a sense of collective ownership over the system's performance, making it clear that sustaining success was a shared responsibility.

Moreover, continuous improvement in an ERP environment is deeply intertwined with the concept of agility. No matter how well the system is calibrated at the outset, market conditions and internal processes are constantly in flux. Thus, it is essential to build a culture that is open to reappraising and adjusting strategies as needed. This requires a proactive risk management component.

Invest in Ongoing Training and Upgrades

In addition to these internal processes, sustaining success calls for an ongoing investment in advanced training and system upgrades. An ERP system serves as a living platform that evolves in tandem with technology and business needs. The training that began during implementation must be refreshed as new features are introduced and as your organization's operational landscape shifts.

In one case at a financial services firm, the initial rollout was followed by a series of update sessions, each focused on new modules and enhancements that had been integrated after the system went live. The firm categorized training by user roles and levels of complexity, ensuring that everyone from junior staff to senior executives remained up to date.

This commitment to continued education prevented the ERP system from becoming static; instead, it enabled the organization to continuously unlock new functionalities, drive efficiency and maintain a competitive edge in an industry where timing and accuracy are paramount.

It is essential to develop a sustainable training framework that includes regular updates, interactive learning modules and mentorship opportunities.

For example, a manufacturing firm I once advised set up a structured continuing education program for its ERP system. Rather than rely on occasional refresher sessions, they implemented an online training portal where employees could access updates, participate in webinars and even earn certifications related to system usage.

This approach enhanced user proficiency and fostered an environment where employees stayed current with technological changes. When workforce competencies improve continuously, the ERP system remains a tool for innovation and efficiency rather than becoming a source of frustration.

Cutting Edge Tech with Business Edge

A related focus is the alignment of embracing cutting-edge technologies that offer a genuine business edge. Adopting something like predictive modeling or machine learning should never be done solely because it seems cutting-edge. Instead, you want to target specific pain points or strategic goals that advanced analytics can address more effectively than conventional reporting.

For example, a manufacturing distributor grappling with high return rates found that standard dashboards only revealed the extent of the problem, not its underlying causes. By integrating a machine learning algorithm into their ERP system, they correlated product returns with aspects like supplier batch data, time of purchase and store location, unveiling patterns that old methods missed. As a result, they discovered that certain product lines had higher return rates specifically when inventory management practices—like temperature control or shelf rotation—were not consistently enforced.

Using this information, the distributor revised inventory-handling procedures, cut return rates significantly and strengthened supplier contracts to minimize future losses. This story highlights that advanced analytics within ERP are the most potent when they address a tangible strategic need, transform raw data into practical knowledge, and lead directly to operational improvements.

Encourage Feedback

Feedback loops are also a cornerstone of a sustained ERP strategy. Establishing mechanisms that allow users to report issues, suggest improvements and even share success stories can significantly enrich the knowledge base surrounding the system, driving iterative enhancement.

One particularly successful example comes from a global manufacturer that implemented an online feedback portal as part of its ERP system. Employees across all levels were encouraged to document challenges and innovative solutions.

Over time, this repository of feedback was analyzed to uncover recurring issues, many of which were resolved by a simple change in process or an upgrade in system functionality.

This methodology enabled employees, gave them a stake in the success of the ERP system, and created a culture where every individual contributes to continuous improvement, reinforcing the idea that the system is truly a living part of the organization.

Equally important in maintaining momentum is the integration of feedback mechanisms within the ERP framework. Technology is only as effective as the insights it creates, and feedback from your staff can illuminate unforeseen challenges that might otherwise impede progress.

In one scenario, a financial services firm installed an ERP system and found that the automated report generation was not meeting the specific needs of its loan officers. Instead of accepting sub-par functionality, the organization implemented a continuous feedback process that allowed end users to submit detailed observations and suggestions.

Over time, this led to significant refinements in the reporting tool, transforming initial dissatisfaction into a system that enhanced decision-making across multiple departments. Feedback loops like these help in resolving immediate concerns and create an environment of continuous improvement, with every stakeholder feeling that their input makes a tangible difference.

Continuous Improvement

Leveraging ERP for competitive advantages is not solely about measurement or external comparisons; it is also about fostering an enduring culture of continuous improvement. The ERP system is not a static installation but a living component of your organizational infrastructure that must evolve with changing business needs. To support this evolution, it is essential to establish robust post-implementation processes that include routine audits, regular training updates and strategic reviews.

A particularly instructive example comes from a distribution enterprise that scheduled quarterly system reviews. During each review, the leadership team examined quantitative data and qualitative feedback from employees. Over time, they identified recurring patterns, such as a consistent lag in data refresh rates or a recurring issue in the user interface.

Armed with these insights, the company worked closely with the vendor to introduce system updates that resolved the issues and improved the overall user experience. This continuous loop of evaluation and refinement ensures that the ERP system remains a dynamic asset, one that adapts well to market changes and internal growth.

By committing to continuous improvement and keeping your performance metrics under regular scrutiny, you ensure that your ERP system remains an integral part of your strategy—a living platform that drives long-term competitive advantage and operational excellence.

Effective and Motivational Leadership

Effective leadership is paramount at every step of the continuous evaluation process. Leaders must champion the ERP system and be role models who are committed to continuous improvement. They must actively participate in review sessions, champion training initiatives and proactively address feedback.

An executive at a multinational conglomerate once made it a point to personally review ERP performance data during quarterly meetings and then follow up with departmental managers to discuss actionable improvements. This hands-on approach energized the entire enterprise and underscored that sustaining success was a top priority across all levels of the organization.

Leadership involvement, in this context, ensures that the ERP system is viewed as a strategic asset, with every improvement driving the organization toward a greater competitive advantage.

Strategic Flexibility

Furthermore, flexibility in strategy is essential. No matter how comprehensive the initial planning is, the business environment is always subject to change. New competitive threats, economic shifts or technological breakthroughs can quickly render previous benchmarks obsolete. Organizations must regularly reassess their ERP strategy through external consultations, internal audits and iterative planning sessions.

A dynamic, flexible approach allows you to maintain the system's performance and to seize emerging opportunities, ensuring that the ERP remains a driver for future growth.

Sustaining success means maintaining a strong alignment between the ERP system and the ever-changing market environment. When new competitors emerge or market conditions shift, the insights created by your ERP system allow you to adjust tactics quickly.

A notable case was a manufacturing company that, through its ERP dashboards, was able to detect a downturn in a particular product line. Prompt action, based on accurate and timely data, allowed them to reallocate resources, adjust inventory levels and rapidly introduce promotions that mitigated the downturn. Such agile responses illustrate that an ERP system is most valuable when it is monitored but actively used to drive ongoing improvements and strategic pivots.

You now understand the importance of a thorough, continuous post-implementation review and the establishment of meaningful KPIs that capture the full impact of your ERP system.

Regular evaluation, robust feedback mechanisms, ongoing training and external benchmarking collectively ensure that your ERP system continues to deliver value long after installation. These practices confirm that the system is delivering as promised and identify areas for further refinement, ensuring that your technology remains competitive and aligned with your strategic goals.

As you reflect on your current metrics and review processes, consider the ways in which continuous improvement can become a core part of your organization's culture.

Building on these observations, several key strategies emerge to sustain success over time, which are:

1. Ensure that you have a well-defined set of KPIs that are revisited on a periodic basis and adjusted as your operational landscape evolves.

2. Incorporate continuous feedback loops that enable every level of the organization—from frontline staff to senior leadership—to contribute to system improvements.

3. Invest in regular training and system updates to keep pace with technological advances and operational changes.

4. Use external benchmarks to create context for your internal success metrics, ensuring that your system's performance is internally optimized and competitive in your industry.

The consistent theme throughout these strategies is adaptation. When monitored attentively and cared for continuously, the ERP system evolves from a fixed set of functions into a dynamic platform fully integrated with your organization's long-term strategy. This doesn't just safeguard your initial investment; it transforms it into an engine for growth and innovation.

I remember a compelling example from a global manufacturer that used its ERP implementation as a springboard for strategic decision-making. Initially, their focus was purely on production efficiency, but as the system matured, they discovered that integrated analytics were revealing new patterns in supply chain performance. Quarterly reviews unveiled inconsistencies in vendor performance and logistic planning that had previously gone unnoticed.

By addressing these issues directly—refining vendor contracts and retooling supply chain management practices—their corrective measures improved operational efficiency and translated into significant cost savings and higher customer satisfaction.

This example clearly highlights that continuous evaluation does more than fix problems. It creates new opportunities for strategic advantage.

Let the practices and examples discussed in this section guide your efforts. By establishing clear metrics, fostering an environment of continuous learning and integrating systematic feedback mechanisms, you lay the groundwork for ongoing innovation and operational excellence.

The next step is to leverage these ongoing improvements to gain a competitive advantage. Here, the discipline of measurement, coupled with a culture of continuous feedback and improvement, transforms your ERP system from a technology solution into an indispensable strategic asset.

Document to Build Expertise

Documentation of experiences and lessons learned to integrate them into a future roadmap is another critical practice. Every performance review, every challenge met and every adjustment made should be recorded systematically. Establish a comprehensive repository of post-mortem reports and action plans following each review cycle. This archive serves as a valuable resource for identifying long-term patterns and informing future upgrades. Over time, these documents become a resource of expertise—a historical repository of what worked well and what did not.

I remember a company that built a detailed "lessons learned" database after its ERP implementation. This database was used in subsequent projects to avoid repeating past mistakes and to continuously refine training and operational processes.

Such documentation ensures that the organization institutionalizes its knowledge and creates a roadmap for continual improvement, which benefits current and future projects.

Practice Lessons Learned

An essential part of maintaining long-term success is practicing lessons learned. Documenting challenges, solutions and unexpected benefits as they arise creates valuable insights and practicing those improves future decisions.

I recall one instance in which a company faced recurring issues with data inconsistencies shortly after implementation. By systematically logging these incidents and their corresponding resolutions, the team was able to identify recurring patterns and introduce process changes that minimized errors.

The result was a more reliable ERP system and enhanced operational procedures from which future teams could benefit. In this way, every setback becomes a valuable lesson that feeds into a culture of continuous improvement, ensuring that the system and the organization evolve together.

In this section, we draw together the diverse threads of insight and experience that have shaped our perspective on ERP systems—from strategic selection and in-depth preparation to the enduring process of measurement, refinement and forward-looking innovation.

Over the course of these chapters, the overarching theme has remained consistent. A well-orchestrated ERP initiative is more than a technical upgrade; it is a comprehensive transformation that reaches to the heart of your organization's operations, culture and competitive clarity.

As we conclude, it is worth reflecting on the broader implications of your ERP process and envisioning a future where continuous adaptation, advanced data insights and a culture of learning define what success can look like in an evolving market.

Empowerment

One of the most essential lessons we have repeatedly returned to is that the real strength of an ERP system lies in its capacity to unify, inform and empower every nook and corner of your organization. Early on, you put in the effort to outline requirements, balance customization and anticipate challenges. Then, through each stage—implementation, managing resistance, training staff, aligning processes—you built a foundation that transcends operational tasks.

As we discussed advanced analytics, cross-functional synergy and external collaborations, it became clear that an ERP system that can continuously adapt can shape your strategic direction. This synergy between daily operations and long-range strategy marks the turning point where your ERP system shifts from being a behind-the-scenes tool to a core driver of corporate advantage.

Leadership and Culture

The thread running through each subchapter highlights another enduring theme, which is, leadership and culture propel change far more effectively than technology alone. While strong architecture, real-time dashboards, and predictive analytics can give you the raw data you need, it is the people in your organization, supported by strong, engaged leadership, who decide how to use that information.

Because the ERP system is inherently holistic, bridging multiple departments and functions, a user-centric culture becomes an essential ingredient. It fosters open communication, encourages collaboration and motivates each team member to seek out strategic opportunities rather than simply handling routine duties.

In such an environment, learning evolves from a discrete, periodic endeavor into a continual practice deeply woven into everyday operations.

Alongside these technical and operational considerations, leadership engagement remains pivotal. The shift from basic usage to strategic exploitation of an ERP system calls for a leadership team that actively champions a data-driven culture, invests in analytics talent and persists in encouraging cross-functional collaboration.

In previous chapters, we explored how leaders who model engagement and accountability create a ripple effect across the organization. When the system moves into advanced stages—incorporating predictive analytics, external vendor integrations or e-commerce expansions—senior leadership must facilitate the resources, training and strategic context for those initiatives.

A well-led ERP environment is not just about producing advanced analytics or integrated modules. It is about ensuring that every user, from department heads to front-line employees, sees real advantages in adopting these enhancements.

Internal and External Collaboration

Consider how your ERP system can drive deeper internal and external collaboration. Internally, teams that never interacted directly in the past, such as finance and marketing, can now share consistent data, discuss common metrics and collaboratively solve problems that cross departmental lines.

A real-time single source of truth fosters synergy because it presents a unified view of customer data, cost structures and inventory details.

For instance, a hospitality group I observed used ERP integration to link their finance and operational teams, enabling them to see how the timing of specific marketing promotions influenced the costs of event planning and staffing. This visibility allowed them to identify profitable promotional windows, cut overhead expenses and devise cross-department strategies that would have been nearly impossible without consolidated data.

Externally, an ERP system can integrate vendors and partners more closely to your supply chain, facilitating streamlined communications, faster order fulfillment and timely inventory updates. These deeper collaborative bonds reinforce trust, reduce errors and let your operation pivot with greater agility when market conditions shift.

As the global economy becomes increasingly interconnected, collaborative networks and shared data ecosystems will take on an even greater role. Future-minded organizations can use their ERP systems to build secure, resilient links with suppliers, logisticians, distributors and even customers.

This larger network of integrated data helps reduce friction and duplication and creates powerful vantage points for strategic collaboration.

For instance, a manufacturing firm might co-design products with trusted suppliers, feeding real-time design constraints, inventory data and market feedback directly into a shared ERP environment.

Or a manufacturer could create improved purchase experiences by synchronizing inventory and order data across multiple channels, from physical stores to e-commerce platforms and emerging social shopping corridors. In all these scenarios, the ERP system acts as the orchestrator, ensuring consistency, visibility and trust in a networked global economy.

New Roles and Positions

Moreover, as your organization becomes more adept at mobilizing ERP-driven insights, staff roles may evolve beyond traditional boundaries. After the system's initial challenges have been addressed, an opportunity arises to rethink the nature of certain positions—are there new roles enabling data analysis across departments, or positions focused on bridging analytics with strategic planning?

In a manufacturing firm, a newly created "data strategist" role brought together domain expertise from production and marketing. This individual leveraged ERP analytics to uncover how product line expansions might influence market perception or production capacity. The synergy between operational know-how and the system's data outputs gave the company a level of strategic agility it had not experienced before.

This example underscores that the ERP's potential is only limited by the structures and mindsets within the organization. As the data environment grows more sophisticated, new lines of responsibility and collaboration can emerge, giving you the flexibility to respond swiftly to strategic shifts.

Challenges are Bound to Persist

Despite this promising outlook, some challenges are bound to persist. Security and privacy concerns will continue to grow, especially as ERP systems integrate artificial intelligence, external data services and extensive customer information.

Government regulations and industry standards will likely evolve in tandem, imposing stricter compliance requirements. Thus, strong governance and a vigilant security posture must remain priorities.

Just as your ERP system has evolved from focusing on operational tasks to becoming a strategic asset, so too should your approach to compliance elevate from a grudging requirement to a well-managed, ongoing discipline that helps you maintain a strong, trusted market presence.

Of course, none of these expansions and aspirations can take root if your organization does not invest in people, training and leadership commitment. The principle applies as much to future developments as it did during the early days of rollout.

Each new module, each advanced analytics component, or each external integration must be accompanied by clear communication, dedicated support and adequate upskilling. This helps your workforce accept new tools and fully understand how to wield them for strategic gains.

When staff view the ERP system as an enabler for engaging, meaningful tasks, and not just an added layer of complexity, they become ambassadors of ongoing progress, fueling the cycle of continuous improvement.

CHAPTER 9

THE FUTURE OF ERP IMPLEMENTATION

CHAPTER 9: THE FUTURE OF ERP IMPLEMENTATION

As your ERP system matures, its value no longer rests solely on more efficient processes or streamlined data management. It becomes a framework for competitive positioning, market insight and strategic foresight.

Mastering ERP implementation requires working with evolving insights, hard-earned lessons and continuous adaptation. Achieving ERP Mastery means being a resilient leader who can effectively apply the lessons learned from years of ERP implementation. A willingness to be challenged and the ability to bring clarity to chaos when the stakes are high are essential for achieving ERP Mastery. That's when one realizes that reaching a certain level of mastery is just a milestone, while indicating the journey ahead will continue the process of evolution.

ERP has long been the backbone of business operations, enabling efficiency, connectivity and strategic insight. However, as technology evolves and business landscapes shift, ERP implementation must adapt to meet the demands of the future. For business leaders, understanding emerging trends in ERP is crucial to ensuring long-term scalability, resilience and innovation.

ERP Solution for Tomorrow's Business

In the near future, your ERP system will become more than a giant record-keeper of what has already happened. It will be a living, thinking, predicting, doing entity—a trusted digital colleague—working quietly behind the scenes to support and augment improvement.

The Rise of Cloud-Based ERP

Traditional on-premises ERP systems are giving way to cloud-based solutions, providing flexibility, scalability, and cost-effectiveness. With cloud ERP, businesses can improve collaboration and reduce IT infrastructure burdens. The future will see an increasing shift toward fully cloud-native ERP environments, making integrations more seamless and updates more automatic.

AI and Automation Transforming ERP

Artificial Intelligence (AI) and automation are revolutionizing ERP systems, driving efficiency through predictive analytics, intelligent automation and self-learning processes. AI-powered ERP solutions will enhance decision-making by identifying patterns, forecasting market trends and automating routine operations. Business leaders will need to harness AI to optimize workflows, improve accuracy and reduce operational costs.

Integration of IoT and Real-Time Data Analytics

The Internet of Things (IoT) is introducing real-time connectivity between billions of devices and business processes. Future ERP systems will integrate IoT sensors to monitor production lines, track supply chains and optimize asset performance. Enhanced real-time data analytics will empower businesses with instant insights, enabling faster decision-making and proactive problem-solving. ERP will integrate with IoT devices to provide real-time data on manufacturing, logistics and inventory management, enhancing operational efficiency and predictive maintenance capabilities.

Enhanced User Experience with Low-Code and No-Code Platforms

ERP implementation is becoming more intuitive with the rise of low-code and no-code platforms, allowing non-technical users to configure workflows, automate processes and customize dashboards. This democratization of ERP will reduce dependency on IT teams, accelerating implementation and driving user adoption.

The Expansion of Blockchain in ERP

Blockchain technology is set to play a significant role in ERP by enhancing security, transparency and trust in business transactions. Future ERP systems may leverage blockchain for supply chain traceability, secure financial transactions and compliance verification, reducing fraud and strengthening data integrity.

ERP for Hybrid and Remote Workforces

The shift toward hybrid and remote work environments requires ERP solutions that provide seamless accessibility from anywhere. Future ERP implementations will prioritize mobile access, remote collaboration tools and decentralized data management, ensuring workforce agility and operational continuity.

Sustainable and ESG-Focused ERP

As businesses prioritize sustainability and Environmental, Social, and Governance (ESG) goals, ERP systems will adapt to track carbon footprints, manage ethical sourcing and ensure compliance with global sustainability standards. Future ERP solutions will help businesses meet environmental commitments while maintaining profitability.

Here are the contemporary trends in ERP evolution.

Autonomous ERP Processes

At the heart of this vision are autonomous ERP processes. Instead of waiting for humans to push buttons, these systems will identify opportunities, predict outcomes and act automatically—like ordering more materials when sensors show low inventory, or rerouting shipments based on weather forecasts—without asking you every step of the way. ERP that thinks first, acts smart and informs proactively.

Smart Bots / AI Agents

Backing these decisions will be AI-driven digital workers—smart bots (agents) embedded directly into your ERP system that manage entire workflows. Need the books closed at quarter-end? A digital accountant is already on it. Need thousands of invoices processed? Done. These AI "co-workers," also known as “AI agents,” will have their own roles, permissions and performance reviews—yes, seriously—integrated into the org chart.

Quantum Computing

Quantum computing implications hint that within a decade, ERP systems might tap into quantum processors to optimize mind-bendingly complex problems, like supply chain logistics across thousands of locations instantly. Quantum ERP won’t just be faster; it'll solve problems that today's computers would take days to compute.

Metaverse and VR

Thanks to the industrial metaverse and VR collaboration, factories, supply chains and entire companies will be rendered as real-time, explorable virtual worlds. In your daily work, your team won’t just look at dashboards, they’ll step inside them. Imagine your maintenance manager walking through a VR replica of your manufacturing plant to check a pressure valve alert triggered by real-world IoT sensors feeding data into the ERP. To view the problem in real-time, no hard hat required, just a VR headset.

Cryptocurrency and Digital Currency

Money itself is transforming. As cryptocurrency and digital currency integration becomes mainstream, ERPs will handle not just dollars and euros but also stablecoins, CBDCs and maybe even your favorite meme coin. Treasury management will mean tracking blockchains, not bank ledgers. Settlements will happen in seconds, globally, with smart contracts auto-triggered by ERP orders.

Wearable Interfaces

You won't even need a laptop to manage it all. Wearable and voice interfaces will turn your ERP into a whispering assistant living on your smartwatch or your smart glasses. A warehouse worker can get an inventory update whispered into their AirPods. A CFO can approve a multi-million-dollar purchase order with a "yes" spoken into a wearable mic.

5G and Ubiquitous Connectivity

All of this will be powered by 5G and ubiquitous connectivity—meaning factories in the mountains, warehouses in the desert and salespeople on bullet trains in rural Asia will all be live in the system, with low-latency, real-time ERP data in hand. The idea of "offline" will vanish.

Robotics and Drones

Robotics and drones integration means your ERP will dispatch autonomous warehouse robots, drones to count stock in the rafters and innovative forklifts that communicate directly with order management systems. When a drone spots a stockout, the ERP will trigger an auto-replenishment order before any human notices.

Decision Analytics

As all this futuristic machinery hums along, decision intelligence and augmented analytics will mean your ERP doesn't just spit out reports anymore—it will highlight anomalies, recommend actions, simulate scenarios and even suggest what to prioritize next. It'll be like having a brilliant, tireless analyst whispering insights in your ear 24/7, deeply woven into every transaction and decision your business makes.

Unified Enterprise Platforms

Unified Enterprise Platforms (UEP) will collapse the old walls between ERP, CRM, HR and supply chain systems into a single environment. One login. One platform. One golden source of truth for everything, from customer requests to asset depreciation schedules. ERP will stop being a back-office tool and become the nervous system of the enterprise—a living, breathing thing that senses, thinks and acts as one.

The organizations that embrace these developments from the foundation of strong processes and continuous improvement will be the ones best positioned to thrive in a marketplace characterized by rapid change and heightened customer expectations.

Business leaders must proactively embrace cloud migration, AI-driven automation, IoT integration, user-friendly platforms, blockchain security, remote accessibility and sustainability-driven functionalities. By staying ahead of these trends, organizations will position themselves for long-term success, ensuring that ERP remains not just an operational necessity but a strategic driver of growth and transformation.

ERP Implementation for Tomorrow's Business

The future of ERP implementation will be driven by agility, intelligence, purpose-built and user-centric design. Traditional rigid approaches will give way to more dynamic, flexible methodologies that prioritize rapid deployment, seamless integration and continuous optimization. Here's how ERP implementation will evolve.

Experiential Selection

Traditional demonstration of ERP Software will soon be a thing of the past. The on-demand experiential demonstration capabilities of an ERP system are on the rise. An ERP shopper can configure, transact and experience the ERP for their own specific set of transactions and then assess their suitability for their specific needs. The selection process will evaluate an ERP in its entirety—a platform-as-a-solution.

Agile and Iterative Implementation

Future ERP methodologies will embrace agile frameworks, allowing businesses to implement ERP in smaller, manageable phases rather than all at once. Waves of implementations, where the first wave meets the core requirements and each subsequent wave delivers added value, will be the north star of every implementation methodology. This iterative approach ensures faster value realization, continuous improvement and adaptability to changing business needs.

AI-Powered Automation

Artificial intelligence and machine learning will optimize implementation by automating data migration, system configuration and workflow customization. AI-driven insights will help organizations identify gaps, recommend improvements and minimize downtime.

Low-Code and No-Code Customization

Future ERP solutions will enable business users, without coding expertise, to configure workflows and automate processes through intuitive drag-and-drop interfaces. This reduces reliance on IT teams and accelerates implementation timelines.

User-Centric Adoption Strategy

Successful ERP implementations will prioritize user experience through personalized interfaces, AI-powered assistants and intuitive dashboards. Organizations will invest in continuous training, engagement and feedback loops to ensure high user adoption.

Modular and Scalable Architectures

Rather than adopting monolithic ERP systems, businesses will implement modular ERP solutions that allow selective functionality adoption. A fusion of the best practices and overarching enterprise functions. This ensures cost-effective rich functionalities tailored to specific business needs while allowing future scalability.

Embedded Cybersecurity and Compliance

Future ERP systems will integrate security and compliance from the ground up, leveraging blockchain for data integrity, AI-driven anomaly detection and automated regulatory adherence to ensure governance and risk management.

Digital Twin Integration

Digital twins—virtual replicas of physical assets and processes—will be embedded into ERP systems, allowing businesses to simulate operations, predict bottlenecks and optimize workflows before real-world execution.

Continuous Optimization and Adaptive Learning

ERP implementation is an evolving process. AI-driven insights will provide ongoing performance optimization, adaptive learning and automated updates to keep systems aligned with business growth.

The future of ERP implementation will be lean, intelligent and responsive to evolving business demands. Organizations that embrace cloud-based, AI-powered, user-centric and modular strategies will drive faster adoption, improved efficiency and long-term success. Business leaders must shift from traditional ERP deployment models to a continuous innovation mindset to unlock the full potential of future ERP solutions.

Your ERP Journey

As you contemplate these futures, it is also valuable to reflect on the legacy of your own specific ERP journey. You have likely encountered hurdles, some purely technological, others cultural, perhaps even moments where timelines slipped or budgets stretched. In context, each of these obstacles was a lesson, forcing your organization to refine data handling protocols, re-evaluate training methods and strengthen communications between departments.

These lessons become institutional knowledge that will serve you well as new developments emerge. In fact, the ultimate success of your ERP system is often built on the foundation of challenges, a testament to your organization's resilience and adaptability.

Looking back at the entire arc of implementing and optimizing your ERP solution, from the opening discussions in the earliest chapters to this culminating moment, you can see a clear trajectory. The ERP system was never meant to be purely operational. Yes, daily tasks have become more efficient, data entry more streamlined and production cycles better structured. However, the breakthrough lies in how the system's integrated data environment enables deeper insights, speeds decision-making and fosters a sense of unity across traditionally siloed teams.

The chapters on human engagement, change management and measurement each pointed out that success is an ongoing process, evolving alongside organizational needs and market changes. Now, with the end of this book at hand, you stand more equipped than ever to harness that dynamic potential.

As we wrap up, consider a few encouragements:

- **Keep the lines of communication open.**
  Whether the issue is a fresh integration challenge or a novel analytics feature, transparent discussions should remain the bedrock of your ERP culture. People at all levels need to understand new developments, share concerns early and contribute insights that shape system improvements.

- **Maintain and revisit your KPIs.** Over time, the key performance indicators you established at the onset will likely need adjustment to reflect refined objectives or expanded functions. Periodic reviews prevent stagnation by creating insights to keep your performance metrics cutting-edge.

- **Cultivate strategic leadership.** Even the most advanced system can unravel if leaders do not maintain a clear, consistent vision that aligns ERP investments with overarching organizational goals. The voices of senior management and department heads must consistently reaffirm the system's strategic priority and remain open to new applications, expansions or major shifts in approach.

- **Emphasize a culture of learning and iteration.** Each new technology wave—be it AI-infused analytics, advanced collaboration tools or global supply chain expansions—presents risks in tandem with fresh opportunities. Preparatory training, iterative pilot deployments and strong feedback loops transform uncharted territory into manageable, evolutionary steps.

- **Stay aligned with external trends.** Watch for shifts in consumer behavior, competitor strategies, regulatory environments and potential disruptions from emerging technologies. Understanding these signals allows you to pivot swiftly, ensuring your ERP system's relevancy. The era of static, multi-decade IT solutions is fading. Flexible, cloud-based systems and agile development approaches are stepping forward as the norm.

Looking ahead, it is conceivable that the divide between ERP systems and daily strategic decision-making will all but vanish. An ideal future may see ERP platforms operate as living, adaptive intelligence centers that blend operational data, advanced analytics, market signals and user feedback into a unified strategic cockpit.

Senior leaders and line managers alike could wield free-flowing dashboards that automatically update them with essential metrics, benchmarks, forecasts and alerts, which help them coordinate decisions that ripple across the organization in real time. At that juncture, the notion that technology is separate from strategy will fade. Technology and strategy will be viewed as an intertwined ecosystem that simultaneously addresses daily tasks and visionary ambitions.

That prospective future, although dazzling, will require the same core tenets you have cultivated through your ERP process thus far, which are, commitment, clarity of purpose, strong training, data discipline and fearless adaptability. Tools and methodologies may evolve, but the underlying principles remain relevant. New modules and expansions are only as successful as the cultural environment that welcomes them.

Remember that the future is not set in stone—disruptive shifts can arise from new competitors, unexpected economic turns or leaps in technology. The structure and mindset you developed through your ERP process—a mindset of continuous improvement, data-driven decisions, cohesive teamwork and agility—will be your greatest defense against, and response to, such uncertainties.

By weaving adaptability into the organization's daily mechanics, you empower your teams to pivot effectively, seizing the advantages hidden in every market disruption.

I encourage you to keep your ERP system at the center of an ongoing dialogue about growth in terms of operational excellence and in shaping new market possibilities. The path you have taken—solemn planning, balanced selection, rigorous change management and consistent measurement—has led you to an opportune vantage point, a place where your organization stands confident in its current capabilities and prepared to venture into emerging possibilities.

Do not squander this foundation by letting your ERP environment slip into routine maintenance. Adopt a posture of proactive exploration, seeking out fresh use cases and advanced configurations that align with strategic goals, whether that's entering a new regional market, diversifying your product lines or absorbing another chain of synergy partners.

The completion of your initial ERP project is not the end but a major milestone in a continued spiral of innovation. The best organizations I've encountered have turned the "launch" or "stable" phases into something far more ambitious—a launchpad for integrated analytics, new product strategies, novel customer experiences and alliances that shape industries.

What stands between you and similar breakthroughs is a commitment to keep pushing boundaries. By approaching your ERP system not just as an IT resource but as the strategic engine of your enterprise, you can realize the full potential of your investment.

My references here to executive leaders' comments include one from a well-known pizza chain, which said, "They are an information technology service company dealing with pizza." Another example comes from a race-track company executive, who stated, "They are in the business of decision making. Every tool they deploy is aimed at improving the decision-making ability of the human behind the wheel."

These comments highlight important considerations for any business leader aiming to strengthen their organization's core competencies. Understand the relationship between the outcomes of your business and its enabling systems. Building a business platform that enables the enterprise to perform at its best is crucial. You are the sculptor looking at a marble of possibilities. This book serves as an essential tool—like a chisel—that can help you shape and develop that marble into a business platform for the future of your enterprise.

I leave you, then, with one last closing perspective. Technology, for all its complexity, relies most fundamentally on people—individuals who learn, adapt, create and lead. The chapters of this book underscore how every successful ERP story is a human story, where well-chosen software meets engaged leadership, open communication and a willingness to transform.

Carry that truth with you. The system will keep evolving, new features will be introduced and the operational environment will shift. Through it all, your commitment to harnessing technology wisely and managing change proactively will define your capacity for sustainable success.

May your ERP process continue to reward you with insights, efficiencies and breakthroughs that strengthen your competitive advantage and drive forward-looking growth. With each challenge surmounted and each improvement gained, you add another layer of resilience to your organization.

As you refine and expand your ERP system's reach, you will find that its ability to streamline operations and fuel strategic decisions will only deepen. I wish you fortitude, creativity and success as you move confidently into the future—empowered, informed and fully ready to capitalize on the opportunities that lie ahead.

Index of Key Concepts and Terms

www.ingramcontent.com/pod-product-compliance
Lightning Source LLC
LaVergne TN
LVHW062237200926
848484LV00006B/1554
* 9 7 9 8 9 9 8 7 9 6 2 0 3 *